Pandemics: Catalysts For Change

By Jushenvir Dhillon

Thank you to my family for their support and a special thank you to my little sister Jaaniya Dhillon

Life is but a Flicker in Time, but the Human Spirit & Knowledge Span Eternity

Dedicated to all those who have perished and to all that have saved countless lives in the darkest of days.

4

Table Of Contents

Introduction..............................7

Chapter 1: The Plague..................11

Chapter 2: Smallpox....................27

Chapter 3: Cholera Pandemics......42

Chapter 4: Spanish Flu................59

Chapter 5: HIV..........................79

Chapter 6: SARS & COVID-19.......98

References.............................128

6

Introduction

Most human infectious diseases and pandemics are zoonotic in origin. Zoonoses are diseases of animal origin that have spread to humans. Almost 16% of all deaths worldwide can be attributed to infectious diseases, and zoonoses account for 60% of known infectious diseases and 75% of emerging infectious diseases, according to the National Institutes of Health.

Zoonoses can be transmitted in many ways, including animal and insect bites, petting, or handling sick animals, and consuming undercooked meat, unpasteurized milk, or contaminated water. The types of pathogens that can be transmitted by animals to humans include bacteria, parasites, fungi, and viruses.

Most of humanity interacts with animals, even those without pets or farms. From ticks to squirrels to rats, many members of the animal kingdom can transmit disease. There is mounting evidence which suggests that the increase of zoonotic diseases is directly linked to humans' increasing interactions with animals, many of which are related to our harvesting of animals and land encroachment.

Some of the most well-known zoonotic diseases include SARS, MERS, Ebola, rabies, and certain forms of influenza. Whether originating in wild animals, as is assumed with COVID-19, or in farmed animals, as is the case with avian and swine flu, they all pose serious threats to individual and global health. Increasing incidents of human-animal interaction, such as high-density farming or consumption of wild animals, as seen in Africa and China, increases the risk of zoonotic diseases. Like wildlife markets, these food production systems are breeding grounds for infectious disease. Packing thousands of domestic livestock together in a cramped space significantly accelerates the spread of zoogenic pathogens. This results in "cross breeding" of different viruses through exchange of genetic material that can make viruses more deadly and infectious. A clear example of this are the globally recurring Influenza epidemics, where exchange of genetic material between the influenza strains of swine and fowl merge to generate new variants of Influenza every year. Therefore, the increased interaction of various animals and the close contact with humans amplifies the rate of viral mutations and the chance of spillover to humans.

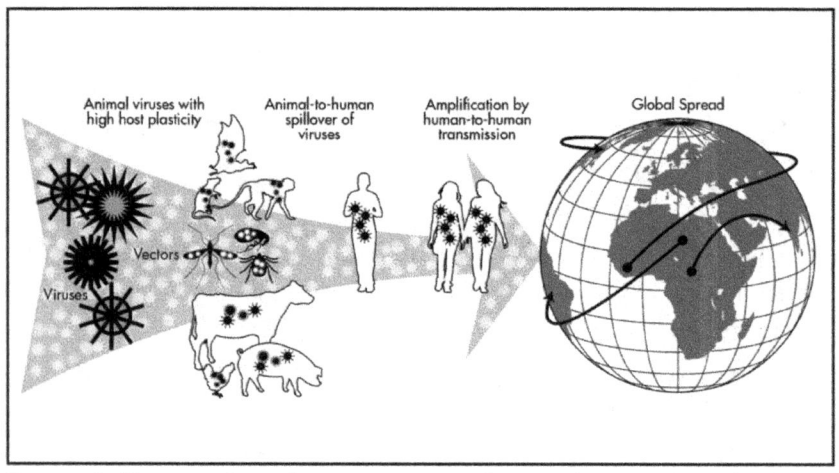

Source: Spillover and pandemic properties of zoonotic viruses with high host plasticity, Johnson et. al

A final major source of zoonotic diseases is habitat destruction from logging, mining, and agriculture. All these practices force many species of animals closer together, facilitating viral transmission and mutations.

International travel and trade have accelerated the spread of pandemic diseases around the globe. There is substantial evidence that outbreaks of animal-borne and other infectious diseases are on the rise. The World Health Organization (WHO)

tracked about 1,500 epidemic events in 172 countries during the period between 2011 and 2018, and it is highly likely that the current coronavirus crisis is only a forewarning of what is yet to come.

Zoonotic diseases are responsible for about 2.5 billion cases of illness and 2.7 million human deaths worldwide, every single year and exceed deaths from traffic accidents (1.24 million deaths) and diabetes (1.37 million deaths) combined globally. Even non-fatal zoonoses cause significant damage to human health, societies, and economies, given that one out of four people on this planet is affected by a zoonotic disease, annually.

When it comes to assessing the hazard posed by a virus, the most widely used measure is the case-fatality rate. While the impact of COVID-19 is unparalleled in modern times, it is not nearly as deadly as some other zoonotic diseases. Although the actual case-fatality rate is still under debate, the case-fatality rate varies greatly according to region, with a current average of 2.2%. This makes COVID-19 substantially more dangerous than regular flu, which has a case-fatality rate of less than 0.1%, and equal to that of the Spanish Flu that killed nearly one-third of the human population. COVID-19 case-fatality rate, however, is dwarfed dramatically by the avian flu with rates of up to 60% (H5N1) and Ebola with a case-fatality rate of 90%. If the case-fatality rate of a future global zoonotic outbreak is similar to those of Ebola, H5N1, or the 1918 flu pandemic, its effects will certainly overwhelm virtually all existing infrastructure. In that scenario, it will not be a question of enough ventilators or intensive care beds, but that of enough doctors and nurses to treat the infirmed.

As will be discussed in the following chapters, pandemics have accelerated changes in society and technology. Each pandemic has resulted in technological advances that have helped to mitigate casualties from pandemics, foster the development of vaccines, give birth to new scientific fields such as epidemiology, lead to scientific discoveries of bacteria and viruses, and many others that will be discussed. These advancements have helped to change the global approach to fighting pandemics and developing strategies to avert the severity of future pandemics.

In addition, pandemics have led to seismic changes in society including extinction of civilizations, geopolitical changes with the rise and fall of empires, fall of feudalism and shift to capitalism, increasing clout of the labor class, increasing wealth gap between citizens and between nations and exposing the deep-seated inequalities within and between nations. As devastating as the loss of millions of lives over the course of human history from infectious diseases, pandemics have catalyzed changes that have accelerated technological and medical advancements and forced societal changes that continue to improve the human condition.

Chapter I

The Plague

Origin

The plague, caused by the bacterium *Yersinia pestis*, was the cause of some of the world's deadliest pandemics, including the Justinian Plague, the Black Death, and the major epidemics that swept through China in the late 1800s. The disease continues to affect populations around the world today. Despite its historical and modern significance, the onset and origin of the disease are not well understood. Exactly when and where *Y. pestis* acquired the virulence profile that allows it to colonize and transmit through the flea vector is not well understood.

Previous studies of ancient *Y. pestis* genetic profile identified its earliest origin dating back to the Late Neolithic (10,000-4,500 B.C.) and Early Bronze Age (3300-1200 B.C.). An international team of researchers led by the Max Planck Institute for the Science of Human History analyzed 3,800-year-old *Y. pestis* genome recovered from two individuals in a double burial site in the Samara region of Russia (Figure 1). Their genetic analyses suggest a Bronze Age origin for bubonic plague.

Figure 1. Source of oldest strain of *Y. pestis* in burial in Samara region of Russia.

This strain is the oldest sequenced to date that contains the virulence factors considered characteristic of the bubonic plague and is ancestral to the strains that caused the Justinian Plague, the Black Death, and the 19th century plague epidemics in China.

Y. pestis can survive in rodent populations and is spread to other mammals, including humans, through flea bites. The point of origin for the Black Death was most likely a population of marmots—small, prairie-dog like rodents—in Central Asia.

Marmots generally avoid contact with humans, but rats will frequently come in contact with both marmot and human populations. Rats also carry fleas, making them an ideal vector for spreading the bubonic plague. In high density areas such as cities, the rat population often exceeds that of its' citizens, thereby rapidly spreading the disease during the plague. It is likely that infectious rats from Asia spread the plague to Europe by stowing away on ships traveling along trade routes.

Bacteriology

The modern history of plague began in 1894, when Alexandre Yersin isolated the causative bacterium in culture and identified it under the microscope and was aptly named *Yersinia pestis*.

Y. pestis is an aerobic, gram-negative coccobacillus in the family *Enterobacteriaceae* (Figure 2). Genetic DNA analysis shows that it diverged from its enteric pathogenic relative, *Yersinia pseudotuberculosis*, up to 6000 years ago. Its primary vector for transmission is the *Xenopsylla cheopis* flea. During the Black Death, the flea was transported by the black rat. The bacteria multiply in infected rodents and block the fleas' alimentary canal, causing the fleas to regurgitate the *Y.*

pestis bacteria into its animal host. In 1898, Dr. Paul-Louis Simond showed that fleas from infected rats could transmit the disease to healthy rats, and Ricardo Jorge in 1927 reported that wild rodents serve as a plague reservoir.

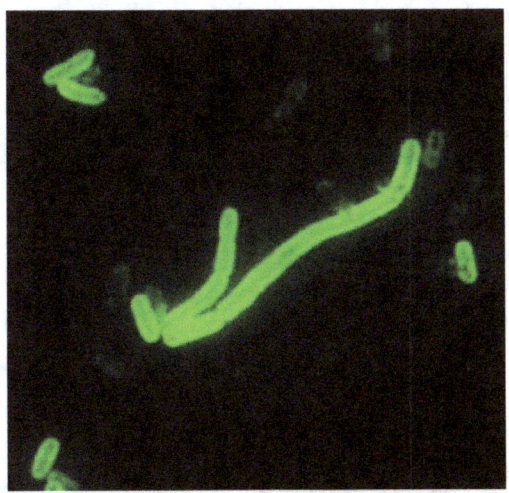

Figure 2. *Yersinia pestis* bacteria.

Three Forms of the Plague

Infection with *Y. pestis* can result in three forms of the plague with different signs and symptoms to those infected:

- Bubonic plague refers to the painful lymph node swellings called "buboes", primarily found around the base of the neck, in the armpits and groin which oozed pus and bled (Figure 3). Victims developed damage to the skin and underlying tissue until they were covered in dark blotches. Most victims died within four to seven days after infection. The bubonic plague was the most common form of plague during the Black Death, with a mortality rate of 30-75% and symptoms including fever of 38 - 41 °C (101-105 °F), headaches, painful aching joints, nausea and vomiting, and

a general feeling of malaise. Of those who contracted the bubonic plague, 4 out of 5 died within eight days.
- The pneumonic plague is an airborne plague that attacks the lungs before the rest of the body. Pneumonic plague was the second most common form during the Black Death with a mortality rate of 90-95%.
- The septicemic plague develops from overwhelming infection leading to seeding of bacteria into the blood stream. The disease is contracted primarily through the bite of an infected insect. Septicemic plague can cause widespread clotting within the body and is almost always fatal. Mortality rate in medieval times was nearly 100%. Septicemic plague is the rarest of the three plague varieties (Figure 4).

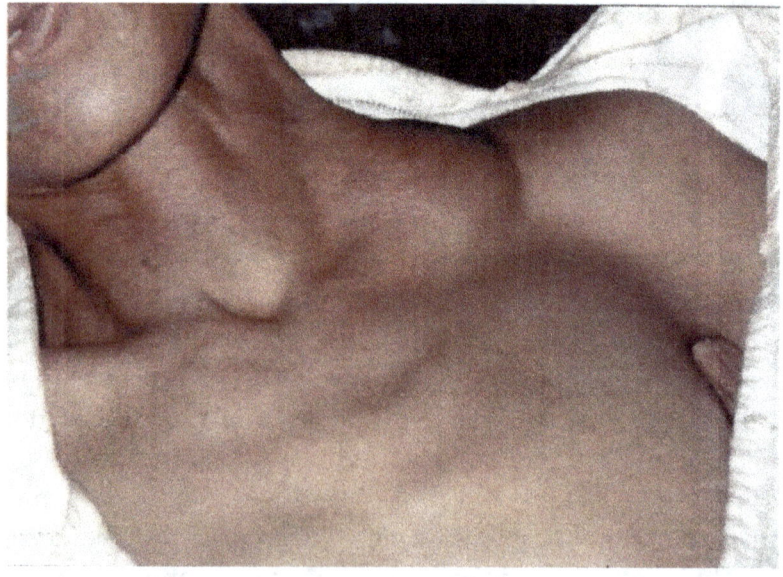

Figure 3. Buboes (swellings). Cervical buboes in a patient with bubonic plague. From Prentice MB, Rahalison L. Plague. *Lancet*. 2007; 369:1196-1207

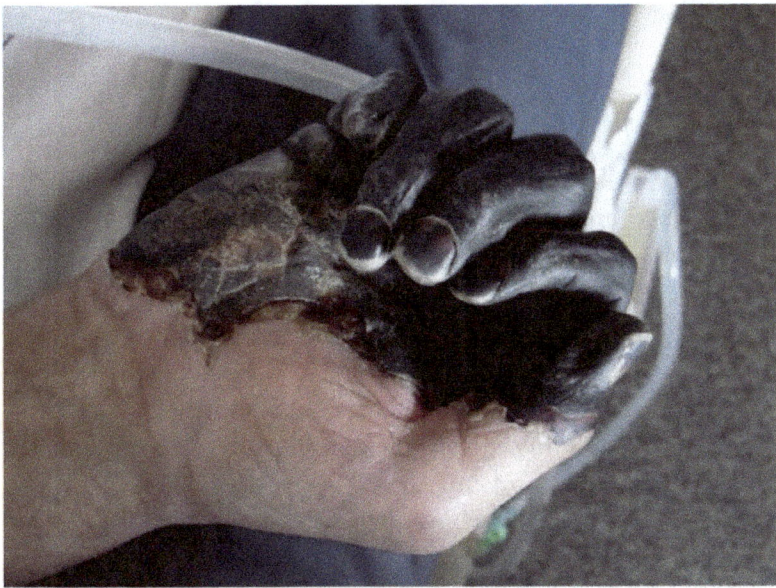

Figure 4. Gangrene from plague sepsis. From CDC.

Epidemiology: The Three Plague Pandemics

There have been three great world pandemics of plague recorded, in 541, 1347, and 1894 CE, each causing devastating mortality of people and animals across nations and continents

The three great plague pandemics had different geographic origins and paths of spread. The Justinian Plague of 541 started in central Africa and spread to Egypt and the Mediterranean. The Black Death of 1347 originated in Asia and spread to the Crimea then Europe and Russia. The third pandemic of 1894, originated in Yunnan, China, and spread to Hong Kong and India, then to the rest of the world.

The Justinian Plague of 541-544

The first recorded pandemic of bubonic plague was the *Justinian Plague* of 541 CE, named after Justinian I, the Roman emperor of the Byzantine Empire at the time. The epidemic originated in Ethiopia, Africa and spread to Pelusium in Egypt in 540. It then spread to the Middle East and was carried on ships on the sea trading routes to the Mediterranean, arriving in Constantinople (now Istanbul) in the autumn of 541.

The Justinian pandemic devastated Constantinople, killing over a third of the city's population. Over the next three years the plague raged through Europe. Between the years 542 and 546, epidemics in Asia, Africa and Europe killed nearly 100 million people.

The pandemic had a drastic effect and permanently changed the social fabric of the Western world. It contributed to the demise of Justinian's reign. Food production was severely disrupted, and an eight-year famine followed. The agrarian system of the empire was restructured to eventually become the feudal system. The social and economic disruption caused by the pandemic marked the end of Roman rule and led to the birth of culturally distinctive societal groups that later formed the nations of medieval Europe.

Further major outbreaks occurred throughout Europe and the Middle East over the next 200 years. The plague continued in intermittent cycles in Europe into the mid-8th century and did not re-emerge as a major epidemic until the 14th century.

The 'Black Death' of Europe in 1347 to 1352

The Black Death of 1347 was the first major European outbreak of the second great plague pandemic that occurred over the 14th to 18th centuries. In 1346 it was known in the European seaports that a plague epidemic was present in Asia. In 1347 the plague was brought to the Crimea from Asia Minor by the Tartar armies of Khan Janibeg, who had laid siege to the town of Kaffa, a Genoese trading town on the shores of the Black Sea. The siege of the Tartars was unsuccessful and before they left, they catapulted over the walls of Kaffa corpses of people who had died from the Black Death. In panic the Genoese traders fled in galleys to Constantinople and across the Mediterranean to Messina, Sicily, where the great pandemic of Europe started. By 1348 it had reached France, Germany, Spain, England and Norway in 1349, and eastern Europe in 1350. The Tartars left Kaffa and carried the plague away with them spreading it further to Russia and India (Figure 5).

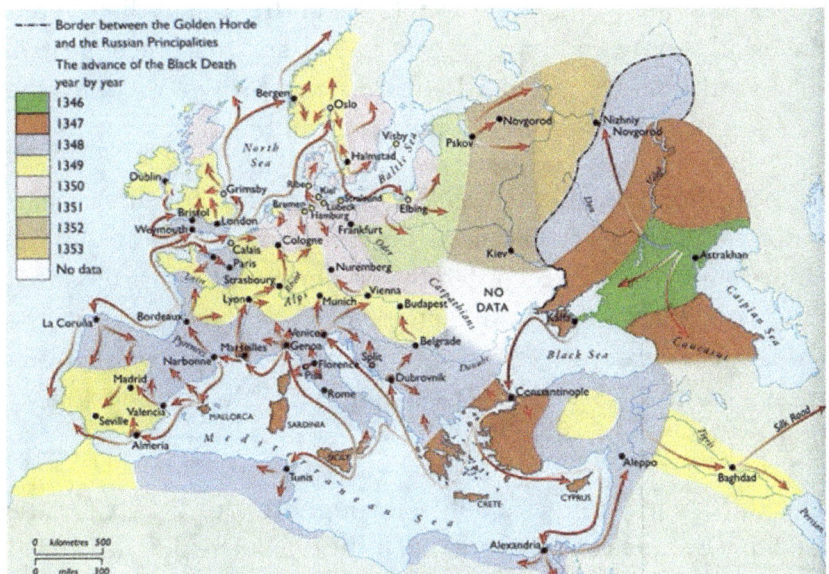

Figure 5. Transmission routes of Black Plague. (Lumen textbook Chapter 8)

In the period 1347 to 1350 the Black Death killed a quarter of the population in Europe (over 25 million people) and another 25 million in Asia and Africa. Mortality was even higher in cities such as Florence, Venice and Paris where more than half succumbed to the plague. A second major epidemic occurred in 1361, the *pestis secunda,* in which 10 to 20% of Europe's population died. By 1430, Europe's population was lower than it had been in 1290 and would not recover to pre-pandemic level until the 16th century.

The Third Pandemic of 1894

The plague re-emerged in the remote Chinese province of Yunnan in 1855. From there the disease advanced throughout China. It had spread to Bombay by 1896 and by 1900 had reached ports on every continent, carried by infected rats travelling the international trade routes on the new steamships. It was in Hong Kong in 1894 that Alexandre Yersin discovered the bacillus now known as *Yersinia pestis,* and in Karachi in 1898 that Paul-Louis Simond discovered the brown rat was the primary host and the rat flea the vector of the disease.

In 1900 the plague came to Australia where the first major outbreak occurred in Sydney. John Ashburton-Thompson, the chief medical officer, recorded the epidemic and confirmed that rats were the source, and their fleas were the vectors in the epidemic. There were 12 major outbreaks of plague in Australia from 1900 to 1925 with 1371 cases and 535 deaths, most cases occurring in Sydney.

The third pandemic circulated throughout the world for the next five decades and did not end until 1959. The plague caused over 15 million deaths, the majority of which were in India.

There have been outbreaks of plague since, such as in China and Tanzania in 1983, Zaire in 1992, and India, Mozambique and Zimbabwe in 1994. In Madagascar in the mid-1990's, a multi-drug resistant strain of the bacillus was identified. Currently around 2,000 cases occur annually, mostly in Africa, Asia and South America, with a global case fatality rate of 5% to 15%.

Fall of Feudalism & Rise of Capitalism

The strongest impact of the plague was in the way it overturned the economic structure prevalent in Europe. Europe in the 14th century was a feudal society, where the king was at the top, and peasant laborers were at the lowest rung of the social ladder (Figure 6). In the middle were the landlords, on whose land the peasants were given the right to live and work. In return, the peasants were expected to pay part of their harvest produce to the landlords as rent. It meant that the landlords could survive on the service of the peasants, while for the peasants (called "serfs") it resulted in never-ending cycles of unpaid work and no hope to rise up the social ladder.

Figure 6. Feudal structure with peasants were indentured servants ("serfs") in exchange for food and shelter. (Wikimedia Commons)

The drastic reduction in the population in Europe after the plague resulted in severe shortage of laborers to work the lands. The shortage of labor catapulted the demand for workers, allowing the serfs leverage for increase wages.

The unprecedented sharp rise in wages became a cause of worry to the landlords who requested the monarchy to intervene. In June 1349, the Crown in England passed the *Ordinance of Laborers*, mandating that all those who do not own lands and are not involved in trade practices be obliged to take up employment offered and accept wages as applicable five or six years ago. The Ordinance also prohibited any landlord from offering higher wages. Despite the order though, wages continued to increase, resulting in the Crown passing a second ordinance in 1351.

Each of these measures failed to contain the newfound economic freedom of the serfs. The lack of labor forced higher wages. Consequently, feudalism in Europe came to an end by the 15th century. The labor class grew stronger and richer and the landlords poorer.

The plague served as a great equalizer and had the greatest impact in reducing the wealth gap due to rising wages of the common person and the biggest decline in the percentage of wealth held by the rich ever seen of any pandemic or war (Figure 7). Market forces took effect and significantly raised the earnings and living standards of the peasants and gave rise to demand-supply capitalism.

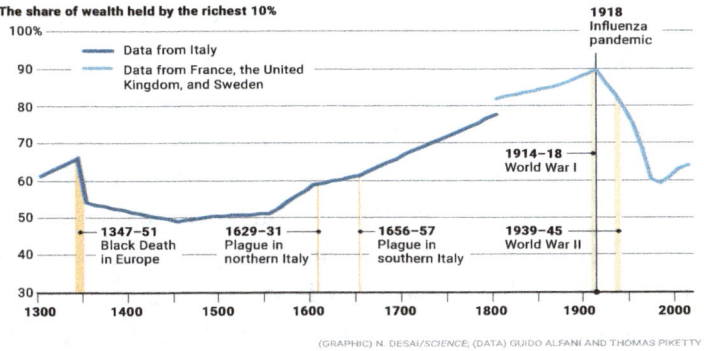

Figure 7. Drop in wealth gap during the plague far exceeded that of any other pandemic or war.

Movement Towards a Modern Age of Medicine

The Black Death propelled medicine past the medieval era and helped to lay the foundations for modern medicine.

The basis of pre-Black Death medicine relied on the theory of humors. Physicians believed the four elements, earth, fire, water, and air, comprised the Earth, and they associated these elements with parts of the body. These false understandings of the body resulted in physicians resorting to their own ineffective methods, such as bloodletting, and inaccurate medical prescriptions to try and aid the sick. After patients and, soon after, physicians died because of these techniques, the Catholic church felt the pressure of the piling deaths. The Pope repealed the previous decree banning surgery as he allowed for the autopsy of dead bodies. It was through these anatomical investigations and dissections that medicine began to take a step forward. Physicians' understanding of human anatomy gave way to improvements in human health and this provided the groundwork for future developments in medicine. This new open-minded attitude towards the discipline allowed for the first post-pandemic generation to make many medical advancements.

The concept of public health arose in Italy following the outbreak of the plague. The rest of Europe followed suit in the formation of public health associations. Members of these associations belonged to the nobility class, but they included medical professionals. These boards developed preventative and counteractive measures to ward off epidemics that included quarantining the infected individuals, collecting information regarding the disease, and then formulating preventative measures.

The interconnection between hospitals and physicians emerged post-pandemic. Pre-bubonic plague, hospitals served as hospice centers that provided sustenance and cleanliness for the infected, similar to hotels. In an effort to move from hospice centers and to take an active role to rehabilitate the sick, hospitals developed relationships with physicians to "attend" to the care of their patients. This facilitated the association between the two that still stands today.

Public Health Measures: Concept of Quarantine

Even in the 14th century the health authorities in northern Italy had established the importance of a 40-day quarantine period, which became the gold standard for continental Europe for the next 300 years. Quarantine derived from the Italian word for forty, "quranta".

The success of the quarantine period to limit the spread of the disease reaffirmed the concept that the plague was a directly infectious disease, and it also showed that it had a long incubation period. European towns gradually realized that the danger in the arrival of an infected and entry was denied if they had come from a town that had suffered an epidemic. Later, in addition to inspecting travelers on arrival, the authorities also required proof that all the towns through which they had journeyed were completely free of plague.

Once an epidemic erupted, those individuals displaying symptoms were removed to emergency primitive isolation hospitals called pest houses (an abbreviation of pestilence). Once a plague case had been identified, the family was locked up in the house, mark of a cross was placed on the door, and a watchman was appointed to stand guard.

Despite only limited medical knowledge at the time, the epidemiology of the plague was fully understood at least by the middle of the 17th century. At that time, they recognized that victims were infectious before the symptoms appeared, the lengthy duration of the incubation period, the necessity of a 40-day quarantine, and the dangers of droplet infection.

Advent of Personal Protective Equipment (PPE)

The first PPE, a head-to-toe protective suit, was invented in 1619 during the great plague in Europe by Dr Charles de Lorme (1584–1678), a physician to the French King Louis XIII. At that time the cause of the bubonic and pulmonary plague was attributed to miasma (a highly unpleasant smell or vapor), bad air, witchcraft, or punishment from God, but everyone could see that the disease spread by contact with the patient or with bodily fluids or cough-related droplets in the air. The protective clothing was meant for plague doctors who tried to "rebalance humors" by bloodletting, or by applying frogs or leeches on the skin lesions. Most plague doctors were not experienced physicians, many even lacked medical training and had little to offer to the dying patients.

A plague doctor's protective clothing featured a light wax-coated leather or fabric overcoat with the shirt tucked into the pants. The boots were also covered. Gloves were attached to the overcoat's sleeves. The original doctor's mask was made from leather coated with wax that resembled the gas masks or gas hoods used in World War I. It had glass eye openings and a small beak that was filled with herbs to purify the putrid air. The plague doctors also carried a cane used to poke or probe the sick person, to avoid close contact. A plague doctor was often the last person the patient saw before they died. They were the harbingers of death (Figure 8).

Figure 8. Plague doctors protective clothing. The mask is preserved in the Deutsches Historisches Museum in Berlin.

Resurgence of Plague in the Modern World

Since plague lives in a variety of rodent species, it's virtually impossible to eradicate. According to the Centers for Disease Control (CDC) in the United States, the bacteria is endemic "in rural areas in central and southern Africa (especially eastern Democratic Republic of Congo, northwestern Uganda, and Madagascar), central Asia and the Indian subcontinent, the northeastern part of South America, and parts of the southwestern United States."

The World Health Organization (WHO) estimates that the three most plague-burdened countries are Madagascar, the Democratic Republic of Congo, and Peru. In 2013, there were 783 cases across the globe and 126 deaths (Figure 9).

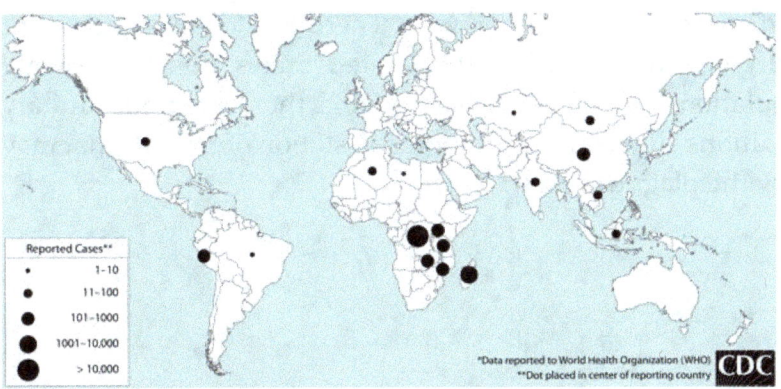

Figure 9. Reported Plague cases by Country, 2000-2009

Wealthy countries, including the US, have mostly purged rat-associated plague from their urban areas. In the United States, plague occurs where there are wild rodents — squirrels, wood rats, prairie dogs, chipmunks, mice, voles, rabbits — mainly in rural areas in the Western part of the country (Figure 10).

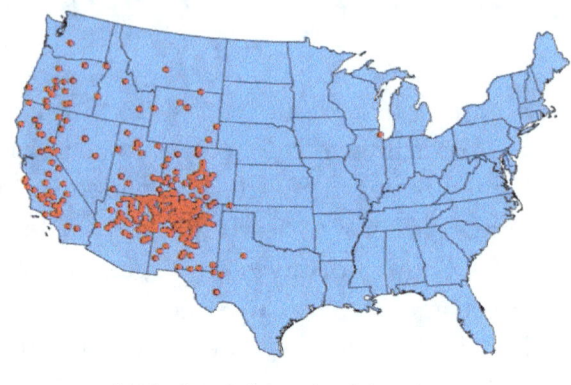

1 dot placed in county of exposure for each plague case

Figure 10. Reported cases of human plague—United States, 1970-2012. Source: CDC

Between 1900 and 2012, there were more than 1,000 cases in the United States. In recent decades in the U.S., plague cases have averaged about seven per year. There is a rising concern for resurgence of plague in the United States due to growing populations of homeless encampments that have poor sanitary conditions and likely to promote infestation of rodents that may carry the plague (*Y. pestis*).

Chapter II

Smallpox

Origin

Smallpox has claimed the lives of millions during its reign of terror. Deaths from smallpox have far outnumbered that of the bubonic plague, the Black Death, and the wars of the 20th century combined. The origin of smallpox remains uncertain but is believed to have originated from livestock 10,000 years ago. Close variants of the smallpox virus are seen in cows, horses and monkeys. Smallpox is believed to have appeared at the time of the first agricultural settlements in northeastern Africa, around 10,000 BC. It is believed to have spread to Egypt from merchants traveling between the two nations. The first suspected case of smallpox was Pharaoh Ramses the V of Egypt in 1157 BC (Figure 1). Findings of pitted marks of the face, shoulders and neck of the mummified remains suggest the possibility that he died of smallpox.

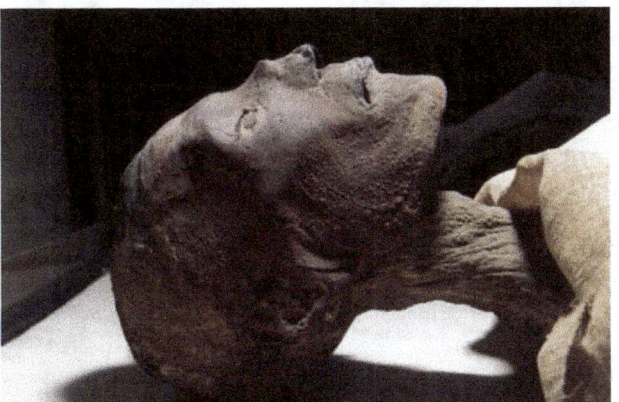

Figure 1. Pharaoh Ramses V with skin lesions akin to smallpox

Smallpox was introduced to Europe sometime between the 5th and 7th centuries with frequent epidemics during the Middle Ages. The disease greatly affected the development of Western civilization. The first stages of the decline of the Roman Empire (108 AD) coincided with a large-scale epidemic (The plague of Antonine) which accounted for the deaths of almost 7 million people including the Roman Emperor Marcus Aurelius. The Arab expansion, the Crusades, and the discovery of the West Indies all contributed to the spread of the disease.

In the New World, smallpox was introduced by the Spanish and Portuguese conquistadors. The disease decimated the local population and was instrumental in the fall of the Aztecs and the Incas empires. Similarly, on the eastern coast of North America, the disease was introduced by the early settlers and led to a decline in the native population. The devastating effects of smallpox also gave rise to one of the first instances of biological warfare. During the French-Indian War (1754–1767), Sir Jeffrey Amherst, the commander of the British forces in North America, suggested the deliberate use of smallpox to diminish the American Indian population hostile to the British. Slave trade was another contributing factor to smallpox in the Americas because many slaves came from regions in Africa where smallpox was endemic.

In the 18th century in Europe, 400,000 people died annually of smallpox, and one third of the survivors went blind. The symptoms of smallpox, or the "speckled monster" as it was known in 18th-century England, appeared suddenly and the sequelae were devastating. The death rate varied from 20% to 60% and left most survivors with disfiguring scars. The death rate in infants was even higher, approaching 80% in London and 98% in Berlin during the late 1800's.

The word *variola* was commonly used for smallpox and was introduced by Bishop Marius of Avenches (near Lausanne, Switzerland) in ad 570. It is derived from the Latin word *varius*, meaning "stained," or from *varus*, meaning "mark on the skin." The term *small pockes* (*pocke* meaning sac) was first used in England at the end of the 15th century to distinguish the disease from syphilis, which was then known as the great pockes.

Virology & Pathogenesis

Smallpox is caused by the variola virus, a DNA virus of the genus Orthopoxvirus (Figure 2). Variola causes a distinctive rash and is often lethal. The name variola comes from the Latin word for "spotted" and refers to the raised bumps that appear on the face and body of infected individuals (Figure 3). Variola is a member of the Poxvirus family of viruses. A close relative of variola within the Poxvirus family, called vaccinia, does not cause smallpox and was used as a vaccine for smallpox. The word vaccine derives from the vaccinia virus.

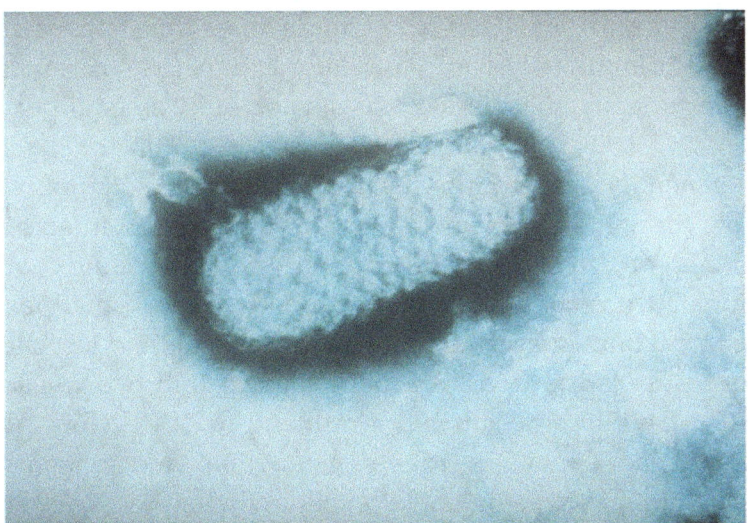

Figure 2. Electron-micrograph of smallpox virus

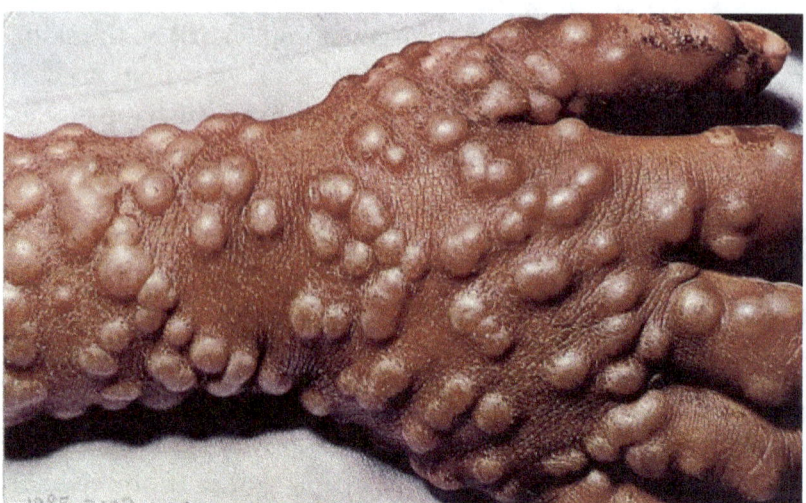
Figure 3. Raised "pockes" on the hand of infected patient with smallpox

A virus related to variola called monkeypox virus made headlines when an outbreak occurred in the Midwestern region of the United States in 2003. This disease, also characterized by a rash and blisters, was the first monkeypox outbreak in the Western Hemisphere. Monkeypox virus sickened about 70 people. The cause was traced to prairie dogs that had been infected by imported African rodents at a pet distribution center. Fortunately, this disease was not as deadly as smallpox.

People generally become infected with the smallpox virus by breathing in virus droplets following exposure to infected individuals or by direct contact with infected fluids or contaminated objects. An unusual property of the smallpox virus is that it only infects humans and not animals and insects. After exposure to the virus, it usually takes one to two weeks before a person becomes ill, and a rash and fever develop. At this point, the person is highly contagious and remains contagious until all scabs fall off after about three weeks. About 30% of infected people die from smallpox infection. People who recover from the infection are often left with permanent scars and sometimes blindness.

The pathogenesis of smallpox involves the following:

- entry for variola virus is usually through the mouth, lungs or skin.
- The virus migrates rapidly to regional lymph nodes.
- Viremia (high number of viral particles in the blood) occurs on the 3rd or 4th day after infection, with further spread of the virus to spleen, bone marrow, and other lymph nodes.
- By the 8th to 12th day after initial infection, infected individuals develop fever and toxemia.
- The virus localizes in small blood vessels of the dermis and oropharyngeal (mouth and throat) mucosa, leading to initial onset of rashes and ulcers in the mouth or throat and skin rashes, at which point (about day 14) the patient becomes infectious. The spleen, lymph nodes, kidneys, liver, bone marrow, and other organs also may contain large amounts of virus.
- Death most commonly results from toxemia (overwhelming immune response resulting in coma and cardiovascular collapse).

Fall of Empires & Tipping of the Balance of Power

A pivotal moment in the shift of balance of power between Europe and the Americas was the arrival of Europeans in the 16th century to the American Continent. The people of the Americas had been isolated from those of Asia and Europe for about 12,000 years. This separation of humanity occurred because at the end of the last ice age, as the world warmed, the ice bridge across the Bering Strait from Asia to North America melted.

Christopher Columbus' arrival in North America created intercontinental connections between Eurasia, Africa, and the Americas that still exist today. It also began a chain of events that dramatically changed the environment, economic systems, and culture across the world. The transfer of goods, people, microbes, and ideas is often referred to as the Columbian Exchange. This exchange created new global networks and radically shaped communities in the Americas.

The Columbian Exchange connected almost the entire world through new networks of trade and exchange (Figure 4). The inter-continental transfer of plants, animals, knowledge, and technology revolutionized the world, as communities interacted with completely new species, tools, and ideas. The Columbian Exchange marked the beginning of a period of rapid cultural change.

After 12,000 years of separation, the Europeans brought upon the Americas a cataclysmic change that forever changed the trajectory of the development of human society and to the development of the modern world. When Christopher Columbus arrived in the Caribbean in 1493, he arrived with 17 ships, 1,500 people, and hundreds of pigs and other animals. Almost all the major species of domesticated livestock were from Eurasia, and the livestock that tend to live closest to humans (cow, sheep, goat, pig, and horse) had been living with Europeans for thousands of years. These provided plentiful opportunities for diseases to pass from animal to human and vice versa, and to spread across Eurasia, from eastern China to western Spain.

Possibly the most dramatic, immediate impact of the Columbian Exchange was the spread of diseases. Prior to contact with the Iberians, indigenous populations thrived across North and South America. There were millions of people (approximately 35 to 75 million) living in the Americas. In the 15th and 16th centuries when the indigenous Americans first encountered Europeans, they were exposed to many diseases including smallpox. Smallpox decimated 90% of the indigenous population and contributed to the fall of Aztec and Inca Empires, paving the way for conquest of the America's by the Europeans.

The deaths of millions of indigenous Americans from diseases introduced by the Europeans caused a labor shortage. Europeans dealt with that problem by forcibly bringing enslaved people from West Africa to the Americas to work on plantations. Over the next few hundred years, more than twelve million enslaved people were brought to the Americas through the Atlantic slave trade system. The Columbian Exchange and its introduction of smallpox completely changed the face of the world. Patterns of production and distribution shifted, as millions of people moved from Afro-Eurasia to the Americas, both willingly and forcibly. Goods—many of which were produced in the Americas by African and indigenous peoples— were distributed around the world. These goods were being circulated in ever-broader networks, creating webs of exchange that shape the world we live in today.

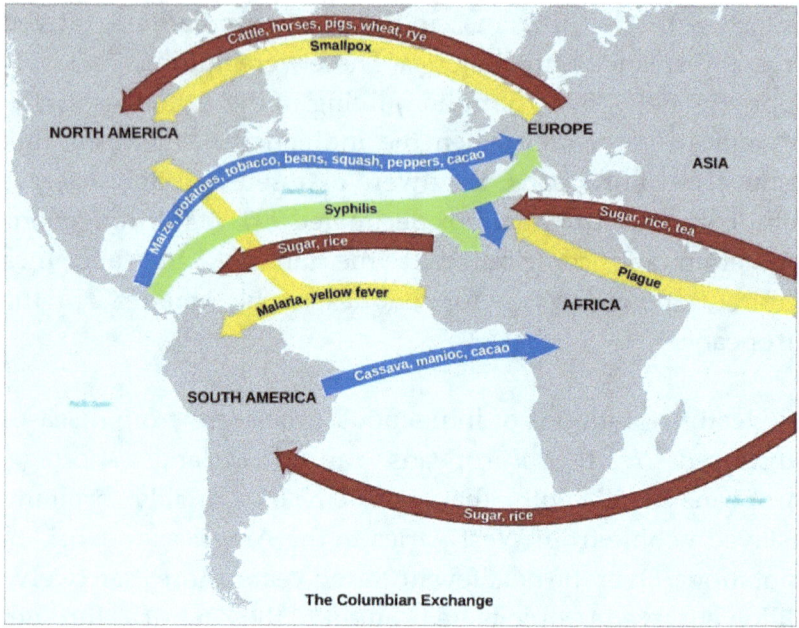

Figure 4.

Introduction of Bio-warfare

Bioterrorism is a primitive form of attack dating back as far as the sixth century BC when the Assyrians poisoned their enemies by contaminating wells with rye ergot, a fungus blight which causes delusions, cardiovascular problems, and death.

Although there was little understanding of germs and the spread of disease in the 14th century, the Mongolian Tartan army hurled the corpses of dead comrades infected with the plague over the walls of the city of Kaffa. Residents of the city fled to Italy, taking the plague with them adding to the second major outbreak of the "Black Death" in Europe.

Some speculate the 15th century Spanish conquistador Francisco Pizarro first introduced smallpox as a biological weapon by offering the Incas gifts of cloth possibly infected with smallpox during his conquest of Peru. Whether the introduction of smallpox was intentional or not, Pizarro was aided in his subjugation of the Incas by the epidemic that raged through the native population.

The first well-documented use of smallpox as a biological weapon was by British troops in the French and Indian Wars (1754-1763). At the time of the Pontiac rebellion in 1763, Sir Jeffrey Amherst, the Commander-in-Chief of the British forces in North America, wrote to Colonel Henry Bouquet: 'Could it not be contrived to send smallpox among these disaffected tribes of Indians? We must use every stratagem in our power to reduce them.' The colonel replied: 'I will try to inoculate the [Native American tribe] with some blankets that may fall in their hands and take care not to get the disease myself.' Sir Jeffrey Amherst, gave the contaminated blankets to Native American, killing as many as half of the population of the infected tribes.

It has been alleged that smallpox was also used as a weapon during the American Revolutionary War (1775-83). During the winter of 1775-76, American forces were attempting to free Quebec from British control. After capturing Montreal, it looked as if they might succeed. But in December 1775, the British fort commander reportedly had civilians immunized against the disease and then deliberately sent out to infect the American troops. A few weeks later a major smallpox epidemic broke out in the American ranks, affecting about half of the 10,000 soldiers. They retreated in chaos after burying their dead in mass graves.

Biomedical Advancement – The Advent of Vaccines

The most successful way of combating smallpox before the discovery of vaccination was inoculation (also called variolation). The word is derived from the Latin *inoculare*, meaning "to graft." Inoculation referred to the subcutaneous instillation of smallpox virus into non-immune individuals. The inoculator usually used a lancet wet with fresh matter taken from a ripe pustule of some person who suffered from smallpox. The material was then subcutaneously introduced into the arms or legs of the non-immune person. The practice of inoculation seems to have arisen independently when people in several countries were faced with the threat of an epidemic. However, inoculation was not without risks. There were concerns those recipients might develop smallpox and spread it to others.

Variolation was likely practiced in Africa, India, and China long before the 18th century, when it was introduced to Europe. In 1670, Circassian traders introduced variolation to the Turkish "Ottoman" Empire. Women from the Caucasus, who were in great demand in the Turkish sultan's harem in Istanbul because of their legendary beauty, were inoculated as children in parts of their bodies where scars would not be seen.

Variolation came to Europe at the beginning of the 18th century with the arrival of travelers from Istanbul. In 1714, the Royal Society of London received a letter from Emanuel Timoni describing the technique of variolation, which he had witnessed in Istanbul.

In Europe, the new methods of variolation became known quickly among physicians. Since there was also a demand for protection against smallpox, physicians soon began the variolation procedure on a massive scale. The death rate associated with variolation was 10 times lower than that associated with naturally occurring smallpox. In the 1750's, more European princes died of smallpox, giving further impetus for the use of variolation.

The regular practice of variolation reached the New World in 1721. Under the guidance of the Rev. Cotton Mather and Dr. Zabdiel Boylston, variolation grew popular in the colonies. When a ship from the West Indies carried people sick with smallpox into Boston in 1721, an epidemic broke out in Boston and other parts of Massachusetts. Boylston immediately started a variolation program and continued to inoculate many volunteers. Mather and Boylston used a statistical approach to compare the mortality rate of natural smallpox infection with that contracted by variolation. During the great epidemic of 1721, approximately half of Boston's 12,000 citizens contracted smallpox. The fatality rate for the naturally contracted disease was 14%, whereas Boylston and Mather reported a mortality rate of only 2% among variolated individuals. This is believed to be the first time that evidence-based analysis was used to evaluate the efficacy of a medical procedure.

In fact, the rapid adoption of variolation in Europe can be directly traced to the efforts of Cotton Mather during the Boston smallpox epidemic in 1721. Although many British physicians remained skeptical even after Mather's success, the data he published was eventually influential. Variolation was subsequently adopted in England and spread from there throughout Western Europe.

In 1757, an 8-year-old boy was inoculated with smallpox, and he was one of thousands of children inoculated that year in England. The procedure was effective, as the boy developed a mild case of smallpox and thereafter remained immune to the disease. His name was Edward Jenner.

Later in his life, Edward Jenner had heard the tales that dairymaids were protected from smallpox naturally after having suffered from cowpox. In May 1796, Edward Jenner found a young dairymaid, Sarah Nelms, who had fresh cowpox lesions on her hands and arms (figure 5). On May 14, 1796, using matter from Nelms' lesions, he inoculated an 8-year-old boy, James Phipps. Subsequently, the boy developed mild fever and discomfort in the axillae. Nine days after the procedure he felt cold and had lost his appetite, but on the next day he was much better. In July 1796, Jenner inoculated the boy again, this time with matter from a fresh smallpox lesion. No disease developed, and Jenner concluded that protection was complete.

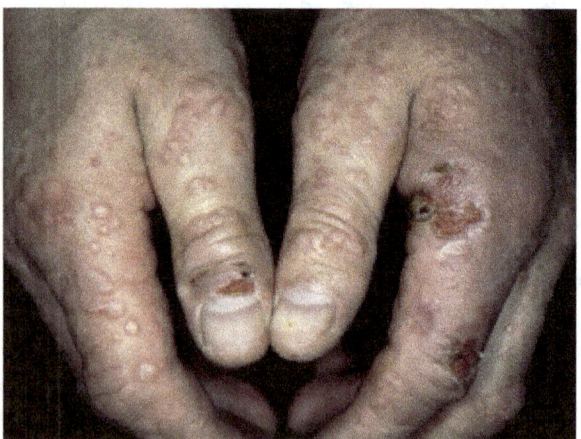

Figure 5. Cowpox lesions

In 1797, Jenner sent a short communication to the Royal Society describing his experiment and observations. However, the paper was rejected. Then in 1798, having added a few more cases to his initial experiment, Jenner privately published a small booklet entitled "*An Inquiry into the Causes and Effects of the Variolae Vaccinae*", also known as Cow Pox. The Latin word for cow is *vacca*, and the scientific name for cowpox is termed *vaccinia*. Based on its Latin derivative, Jenner decided to call this new procedure 'vaccination' whereby an attenuated version of a virus is used to produce a protective immune response.

Jenner's work represented the first scientific attempt to control an infectious disease by the deliberate use of vaccination. He did not discover vaccination but was the first person to confer scientific validity to the procedure and to pursue its scientific investigation. Benjamin Jesty (1737–1816) has been recognized as the first to vaccinate against smallpox. When smallpox was present in Jesty's locality in 1774, he was determined to protect the life of his family. Jesty used material from udders of cattle that he knew had cowpox and transferred the material with a small lancet to the arms of his wife and two boys. The trio of vaccinees remained free of smallpox, although they were exposed on numerous occasions in later life. Benjamin Jesty was neither the first nor the last to experiment with vaccination. In fact, the use of smallpox and cowpox was widely known among the country physicians in the dairy counties of 18th-century England.

Since Jenner, vaccines against many diseases have been developed to confer protective immunity that results after natural infections. As a result of vaccine development, millions of lives have been saved. The power of vaccines is also being applied not only to prevent but also to treat diseases, such as cancer and AIDS.

International Cooperation b/w Nations

In 1958, a delegate of the USSR to the WHO (World Health Organization) proposed that global smallpox eradication be undertaken by the WHO. The 12th World Health Assembly unanimously approved the program in 1959 and embarked on funding the development of a vaccine. In 1967, they proposed a 10-year goal to achieve eradication. At that time 10 to 15 million cases of smallpox were occurring annually in 31 countries with a total population of 1 billion. The strategy of the program was to vaccinate 80% of the population to reach heard immunity and to establish systems of surveillance and containment of outbreaks. The process of worldwide eradication of smallpox was set in motion.

With these measures and with the technological breakthrough of a heat-stable freeze-dried vaccine, the global smallpox incidence began to decline rapidly. The importance of the development of a vaccine that did not require refrigeration cannot be understated. Many of the countries most heavily infected with smallpox had unreliable means of refrigeration to store the vaccine and freeze-drying assured potency and ease of storage and dissemination. The earliest success occurred in 21 African countries that had some of the highest cases of smallpox. They became smallpox free in three and a half years after the initiation of the eradication program.

There were setbacks with resurgence of smallpox in countries like Bangladesh, Botswana, South Africa, and Iran that were thought to have eradicated the disease. In September 1973, new surveillance tactics were developed to address re-infections with tens of thousands of permanent and temporary health workers deployed in house-to-house searches and containment. By late 1975 smallpox was eradicated in Pakistan and in 19 other Asian countries. As cases arose, personnel and resources were mobilized to those countries to assure containment. The last outbreak was last recorded on October 26,1977. The last case of smallpox occurred 10 years, 9 months and 26 days after the program began.

Chapter III

Cholera Pandemics

Origins of Cholera & The Seven Cholera Pandemics

Cholera was endemic in Asia until 1817, when the first pandemic spread from India to several other regions of the world (Figure 1). This pandemic emerged during a period of increasing globalization stemming from technological progress in transportation. The advent of steamships and railways allowed for a dramatic increase in travel and a rise in trade. At that time, health prevention strategies were essentially the same as those implemented during the Black Death. Infected persons were isolated in lazarettos (quarantine stations housing maritime travelers). Port entry was forbidden for ships arriving from regions where cholera was present. Travelers who encountered infected persons or who came from a place where cholera was circulating were quarantined.

Years	Pandemics	Pathogens	Vectors
541–543	Plague of Justinian	*Yersinia pestis*	Fleas associated to wild rodents
1347–1351	Black Death	*Yersinia pestis*	Fleas associated to wild rodents
1817–1824	First cholera pandemic	*Vibrio cholerae*	Contaminated water
1827–1835	Second cholera pandemic	*Vibrio cholerae*	Contaminated water
1839–1856	Third cholera pandemic	*Vibrio cholerae*	Contaminated water
1863–1875	Fourth cholera pandemic	*Vibrio cholerae*	Contaminated water
1881–1886	Fifth cholera pandemic	*Vibrio cholerae*	Contaminated water
1885–ongoing	Third plague	*Yersinia pestis*	Fleas associated to wild rodents
1889–1893	Russian flu	Influenza A/H3N8?	Avian?
1899–1923	Sixth cholera pandemic	*Vibrio cholerae*	Contaminated water
1918–1919	Spanish flu	Influenza A/H1N1	Avian
1957–1959	Asian flu	Influenza A/H2N2	Avian
1961–ongoing	Seventh cholera pandemic	*Vibrio cholerae*	Contaminated water
1968–1970	Hong Kong flu	Influenza A/H3N2	Avian
2002–2003	Severe acute respiratory syndrome (SARS)	SARS-CoV	Bats, palm civets
2009–2010	Swine flu	Influenza A/H1N1	Pigs
2015–ongoing	Middle East respiratory syndrome (MERS)	MERS-CoV	Bats, dromedary camels
2019–ongoing	COVID-19	SARS-CoV-2	Bats, pangolins?

Figure 1.

First: 1817-1824

The first known cholera pandemic emerged from the Ganges River in Bengal, India. Cholera spread to nearly every region of India via trade routes and wartime expeditions. Cholera killed hundreds of thousands of Indians and thousands of British troops. The disease extended to China, Indonesia, and even the Caspian Sea before it finally receded during the winter of 1824.

Second: 1829-1837

Although the true origin is not fully known, the second cholera pandemic struck the European continent. Russia, Hungary, and Germany suffered hundreds of thousands of deaths due to the disease, and the affliction even reached Egypt where it produced similar numbers in death tolls. Additionally, France, England, and even the United States suffered from cholera as well. During this outbreak, transportation was the cause of the rapid spread, as opposed to trade.

With a larger number of developed nations facing the issue of cholera during the second outbreak, the scientific community began to investigate the illness. Many theories about the cause were formed. Scientists in France believed in a correlation between poverty and cholera. Russian scientists concluded that the illness was contagious. The United States blamed the Irish for the arrival of the disease, and some in Britain attributed the cause to be a form of divine punishment.

The most important result of this pandemic was that many nations recognized the disease as an issue and began a concerted effort to eliminate it.

Third: 1846-1860

The third cholera pandemic, like the first, emanated from India and is considered to be the deadliest of them all. From Bengal, India, the bacteria spread to the Middle East. It devastated Russia, where it killed over a million people. Cholera spread to several other nations, where its death tolls reached to 50,000 in England, 15,000 in France, and 150,000 in the United States.

Of the 150,000 Americans that died, it included James K. Polk, the president of the United States at the time. Just as in the 2nd pandemic, many Americans blamed the Irish immigrants for the spread of cholera to the United States. The death of James Polk only worsened the treatment towards the Irish.

In 1852, cholera reached Indonesia and from there it spread to Japan, then the Philippines, and finally Korea. It took lives of hundreds of thousands of people on the Asian continent.

In London, physician John Snow was able to identify the source of cholera. He traced the source of the disease in London back to a contaminated neighborhood water pump (Figure 2). Conducting a study, he proved that contaminated water from the pump was the main agent in the spread of cholera. By the fourth outbreak, London was able to combat the disease because of Snow's findings demonstrating the transmissibility of cholera through contaminated water supplies. Decades later, this conclusion would help prevent future cholera outbreaks in not only England, but also in many other nations. His methodology also laid the foundation for the future development of the field of epidemiology.

Figure 2. The Broad Street pump in London, England memorializing John Snow for his study of contaminated water as a likely source of cholera during the 1854 Cholera Outbreak.

Fourth: 1863-1875

Like the previous cholera pandemics, the Ganges River was the origin of the fourth pandemic. From its origin in Bangladesh, cholera spread via trade to throughout the Middle East and even to Russia, Europe, Africa, and North America. This outbreak killed about 165,000 people in North Africa, 90,000 in Russia, and 115,000 in Germany.

By the fourth cholera pandemic, an understanding of the contagious nature of the disease helped combat this pandemic. In London, a local outbreak prompted epidemiologist William Farr to try and identify the source of the disease. He determined that the East London Water Company was the origin point, and the drinking water from this company was contaminated. Delivering this news to local officials, he was able to prevent further deaths.

These findings by Snow and Farr were not yet made in other nations, as Italy and North America still lacked adequate waterway systems to ensure sanitary conditions.

Fifth: 1881-1896

The fifth cholera outbreak was the last for Europe. William Farr's and John Snow's discoveries propelled defense mechanisms against the disease. Cholera was not seen in Europe again as a result of these medical advancements.

The same was not true for other nations who still failed to understand the contagious nature of cholera. In Egypt, 58,000 people died. Japan lost 90,000 individuals. Spain also lost 120,000 individuals.

Sixth: 1899-1923

Europe's advances in public health and in-depth understanding of the nature of cholera prevented further outbreaks in all of Europe.

The United States experienced its last outbreak with the sixth cholera outbreak. A steamship from Naples, Italy brought the disease to New York City. Isolations and quarantines were then used to help prevent the spread of the disease, and these methods were largely effective. Only eleven people died, far less than the thousands before in prior cholera pandemics.

The sixth cholera pandemic had little effect in western Europe because of advances in public health, but major Russian cities and the Ottoman Empire suffered significant deaths from cholera. More than 500,000 people died of cholera in Russia from 1900 to 1925, exacerbated by the social revolution and warfare taking place at that time.

Seventh: 1961-1975

The last cholera outbreak began in Indonesia, and it was known as El Tor. It reached Pakistan, India, the Soviet Union, and North Africa. The disease emerged in smaller nations such as Jerusalem and smaller districts of Indonesia and Istanbul. The death toll of the last outbreak was far less than previous outbreaks.

Bacteriology: Discovery of Vibrio Cholerae

Two men are credited for having discovered the microbe that causes cholera - *Vibrio cholerae* (Figure 3). Originally it was isolated by Filippo Pacini in 1854, though it was not well known until Dr. Robert Koch brought attention to the disease many years later. It is a gram-negative bacteria found in dirty brackish water.

The earliest original discovery of the comma-shaped *Vibrio cholerae bacillus* goes to Italian anatomist Filippo Pacini for his 1854 paper, "Microscopic observation and pathological deductions on cholera". Pacini discovered the bacteria after a bad outbreak of the disease in Italy that year (Figure 3). Yet, his work went completely unnoticed for a considerable time after his death.

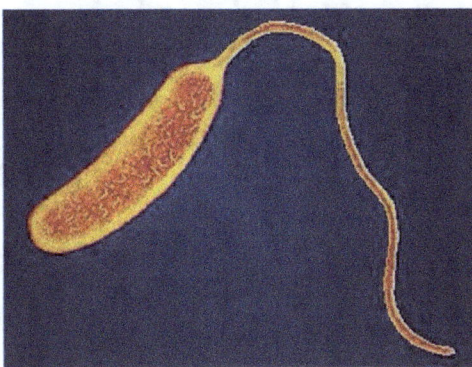

Figure 3. *Vibrio cholerae*

Three decades later, unaware of the fact that Pacini had already done the same research at the University of Florence, Dr. Robert Koch, one of the "fathers" of microbiology and bacteriology, was also researching the microorganism causing cholera. In his quest for the cholera-causing bacterium, he traveled to Alexandria, in Egypt, where an epidemic of the disease was running rampant. Once in Egypt, he performed analyses on the bodies of the victims of the disease and he found a bacillus bacterium, the same one that Pacini had found, in the intestinal lining of cholera victims.

Dr. Koch moved his research to India, which was another cholera "hotspot". There he was able to grow the bacteria in a pure culture. The first thing that Koch noticed was the bent shape of the bacillus *Vibrio cholerae*. He observed that damp earth and moist soiled linen were the places where the bacteria would grow and proliferate. Koch also observed that these organisms were present in great quantity in the "rice water stool" in patients suffering from an advanced state of cholera, but they were never present in patients who suffered from diarrhea that resulted from other causes.

These observations and discoveries that Dr. Koch performed led him to becoming credited for the discovery and isolation of *Vibrio cholerae*. In 1965, 82 years after his death, the International Committee on nomenclature decided to officially name the organism which causes cholera *Vibrio Cholerae Pacini 1854*, in honor of Pacini who was the first one to ever isolate and identify the bacillus.

In 1959, the discovery was made that *Vibrio cholerae* causes diarrhea by producing an enterotoxin. The enterotoxin directly promotes the secretion of fluid and electrolytes from the intestinal lining resulting in severe dehydration from "rice water stools" (Figure 4).

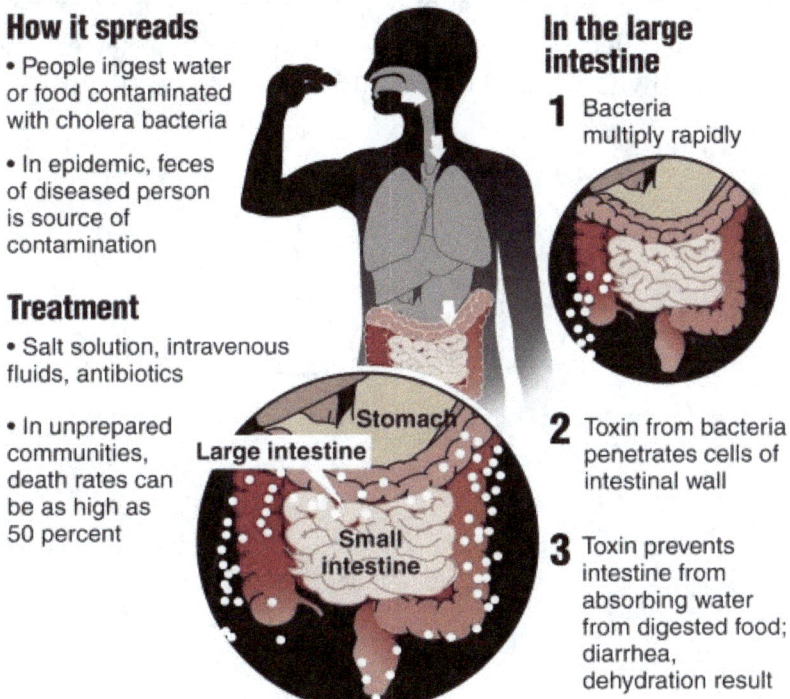

Figure 4. Source: WHO

Transformation of Cities: Development of Sewage System and Municipal Water Supply

In the summer of 1858 in London, England, raw sewage overflowed into the River Thames and caused the "Great Stink." The odor so repugnant that it forced the closure of the Houses of Parliament and the construction of a modern sewer system that transported the city's waste far enough away from

London that the river's tides took it out to sea. Civil Engineer, Joseph Bazalgette was hired and developed the intricate sewer system that was hailed as an extraordinary feat of engineering. It involved the construction of major new 'intercepting sewers' that would collect sewage from the existing array of sewers and moved it down river and eventually into the ocean (Figure 5).

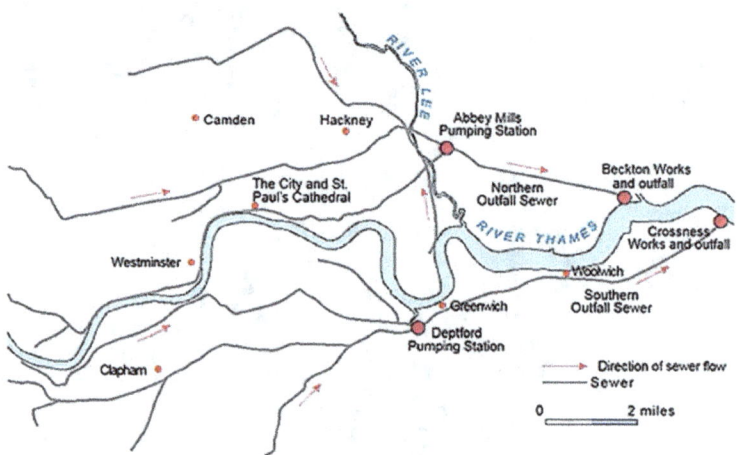

Figure 5. Map of Bazalgette's 1858 sewer system

Most of the sewage system was very carefully designed to flow downstream by gravity, but at critical points, all the sewage would have to be pumped up to a higher water level using huge steam pumps.

One of these pumping stations was at Abbey Mills, where much of the sewage from London's East End had to be pumped up into the enormous Northern Outfall Sewer which flowed across the marshland towards the ocean. This was a prestige project and the buildings above ground, such as the pumping stations, were lavishly decorated (Figure 6).

Figure 6. Restored interior of pumping station in Crossness, London

Joseph Bazalgette replaced 165 miles of old sewers as well as constructing 1100 miles of new ones. He was knighted for his development of the modern sewer system of London in 1854. His work saved the lives of hundreds of thousands of Londoners with prevention of not only Cholera but other forms of disease like typhoid.

In France, Emperor Napoleon III came to power in 1848 amid a cholera outbreak that took the lives of approximately 19,000 Parisians. An admirer of the parks and garden squares of London, the nephew of Napoleon Bonaparte sought to remake Paris in the wake of the pandemic. "Let us open new streets, make the working-class quarters, which lack air and light, more healthy, and let the beneficial sunlight reach everywhere within our walls," he declared. Under the direction of Baron Georges-Eugène Haussmann, French authorities tore down 12,000 buildings, built tree-lined boulevards and parks, erected fountains and installed an elaborate sewage system that transformed Paris into the modern-day "City of Light."

In 1850, Napoleon III and Baron Haussmann hired Eugene Belgrand, an engineer, to design and oversee the building of new sewers. By 1878, Eugene Belgrand had built a 400-mile-long new sewer system with one network for drinking water, one network for water for street cleaning, one network for sanitary sewers, and one network for wastewater.

The sewers consisted of large, six-feet high drains made of sandstone and had roads that sewer workers could easily walk on to facilitate sewer-cleaning. Sewer workers used boats to move the sludge in the sewers to barges that then took and deposited it elsewhere. Sometimes the sludge was also removed via manholes.

Belgrand also built a treatment plant to treat the waste and built aqueducts that brought clean drinking water to the Parisian people. From 1880 to 1914, more than half of the Parisian buildings were connected to the sewers.

There was a dramatic drop in typhoid and cholera cases in Paris because of Belgrand's new sewer system. The French authorities continued to build this sewer system until 1930. At this point, nearly every Parisian street had a sewer under it. The waste from the sewers was taken to the newly built treatment plant for industrial sewage treatment.

The United States also instituted the same urbanization plans with intensive efforts to incorporate municipal water and sewer systems within cities. Historian Charles Rosenberg noted that the defeat of cholera in the U.S. was a result of the country's "fitful jolt into modernity". Cholera thus provided a stimulus for cities, states and federal authorities in the U.S. to build municipal water systems and improve public health infrastructure. In Mexico and Chile and a few other countries in Latin America, cholera prompted a similar impetus to modernize public health systems. The development of a municipal sewer system and bacteria free city drinking water prevented further outbreaks of cholera in developed nations but continues to recur in countries that lack basic sewage and municipal water systems such as Haiti and India today.

Intravenous & Oral Rehydration Therapy: A Legacy of Cholera

Cholera kills by causing excessive loss of body fluids from watery stools, resulting in severe dehydration and subsequent death. It is an ancient disease, and it took many decades of research to develop an understanding into the fundamental biological processes and to establish effective treatments that are now within reach of even resource-poor, underserved regions of the globe.

A more complete understanding of cholera began when physician-scientists undertook clinical studies in South and Southeast Asia where cholera was endemic. Rudolph Virchow, often considered the father of modern pathology, insisted that *Vibrio cholerae* irreversibly damaged intestinal epithelium (lining), causing the loss of protein-rich body fluids. His theory was based on poorly collected intestinal specimens from patients in India suffering cholera that revealed cellular damage to the intestinal lining. Virchow's ideas took nearly a century to disprove. From 1960 to 1970, research by clinician-scientists on site in Dhaka, Bangladesh and Calcutta, India demonstrated that the digestive system was not permanently damaged during cholera infection and with proper therapy, the intestinal lining could heal and absorb electrolyte enriched water. This discovery demonstrated that fluid and electrolyte absorption could be achieved in cholera patients when glucose was added to electrolyte solutions. This formed the foundation for highly effective intravenous rehydration and for oral rehydration therapy (ORT), the two commonly used methods of combating cholera.

Clinical observations by physicians caring for patients during the 1832 cholera epidemic in London led to the first recorded use of intravenous rehydration therapy. However, it was not until 1958, when a U.S. Naval Medical Research Unit in Bangkok, Thailand recorded precise measurements of the volume and composition of stool samples from cholera patients, that a basis was established for an accurately constituted replacement solution. From 1962 to 1964, Johns Hopkins University researchers at the Infectious Diseases Hospital of Calcutta and the Calcutta School of Tropical Medicine modified and simplified intravenous replacement methods. The studies demonstrated that cholera patients, even those in deep shock, could be saved by the rapid infusion of intravenous replacement solutions. Adequate intravenous rehydration of cholera patients in Dhaka and Calcutta reduced their mortality rate from approximately 30–40% to less than 0.5%.

However, intravenous fluid therapy is costly, requires specialized equipment, and must be administered by trained doctors and technicians. It remains inaccessible to individuals in resource-poor regions where cholera and other infectious diarrheal outbreaks are more common. Physicians recognized the need for a simpler form of treatment to replace the water and salt lost by cholera if many more lives were to be saved in developing nations. This led to the development of oral rehydration therapy (ORT) that included electrolytes and glucose to increase water absorption in Cholera afflicted patients.

The simplicity of ORT is that it does not require medical skills to administer, and it is inexpensive. Both of these features make it accessible to nearly everyone, regardless of location or financial resources. Today, over 100 countries have programs to deliver this life-saving treatment, and ORT is used in over 50% of all cases of diarrhea. The global application of ORT has decreased the death rates from diarrheal diseases by more than half in the last 30 years.

Oral Vaccine Development

Dr. Koch, one of the founders of the bacterial pathogen *Vibrio cholerae*, made a discovery that would spark vaccine development. He observed that people infected with cholera became protected from the disease during that same outbreak. The research of Dr. Koch connected the bacteria to infected water supplies. This showed the immense need for clean water in local communities.

In 1885, Spanish physician Jaime Ferrán, who studied under Louis Pasteur, became the first to create a cholera vaccine. He did so after culturing *Vibrio cholerae* and working with the live bacteria. Ferrán became the first to do a mass-vaccination. He used the vaccine to help 50,000 people in Spain during a major cholera epidemic. Later, he also created vaccinations for plague, tetanus, typhus and tuberculosis.

Two scientists, Sawtschenko and Sabolotny, experimented with a killed cholera bacteria "broth" in 1893. The vaccine proved successful in preventing cholera but required too many high doses to be preventive.

Further research and trials continued later in India during the early 1900s. Dried *Vibrio cholerae* tablets were the focus but were not as successful as previous vaccines. The cost and difficulty in their preparation was prohibitive for mass utilization.

In 1959, a microbiologist in Calcutta, India by the name of S.N. De showed that bacteria-free filtrates of *V. cholera* can cause diarrhea. He correctly postulated that *V. cholerae* produced a potent exotoxin that was the cause of diarrhea and not *Vibrio cholerae* bacteria itself. His observations attracted a young American microbiologist, Richard Finkelstein, who went to Calcutta, worked in De's laboratory, and ultimately isolated the toxin. The discovery of the exotoxin produced by *Vibrio cholerae* facilitated the development of present-day oral vaccines to treat patients with cholera.

Currently, there are three oral vaccines available globally: Dukoral, Sanchol, Vachora. The vaccines work by introducing very small amounts of dead cholera bacteria and the nontoxic components of cholera toxin into the intestinal tract. This allows the immune system to make antibodies against the bacteria and the exotoxin.

The vaccines are 65%-90% protective against cholera for at least 4 months after vaccination to up to 5 years after vaccination, with Vachora having the highest efficacy of 90%. Because vaccine effectiveness is not 100% and only offer short-term immunity, cholera vaccines are used mainly for outbreak control and emergency use, rather than for routine vaccination.

The legacy of cholera are the scientific advances in our understanding of the mechanism of action of cholera and revealing that all forms of secretory diarrhea, regardless of cause, can be treated with ORT. This scientific finding has saved the lives of millions of people afflicted with infectious diarrhea.

Cholera Today

Cholera cannot be eradicated as it is a natural inhabitant of aquatic ecosystems. *V. cholerae* serogroups O1 and O139 are responsible for cholera outbreaks in developing countries. In 2019, 923,037 cases and 1,911 deaths were reported to the WHO in 31 countries (World Health Organization, 2020). Between 2008 and 2012, the number of cases was shown to range between 1.3 and 4 million per year with 95,000 deaths globally. The persistence of cholera is related to poor living conditions including shortage of safe drinking water, insufficient sanitation, crowded housing, and the lack of efficient sewage systems. Re-emergence of the disease can also occur following natural disasters such as earthquakes that disrupt access to safe water supply.

Cholera can be prevented by implementation of public health measures to ensure adequate sanitation and safe water supply. The access to safe drinking water and sanitation are among the primary priorities of The World Bank through funding of the Millennium Development Goals and by United Nations through the Sustainable Development Goals. Furthermore, the Water, Sanitation and Hygiene (WASH) program launched by the WHO is central to the prevention of cholera transmission.

CDC's global WASH work is focused in six areas and involves partnerships with other US government agencies, Ministries of Health, non-governmental agencies, and various international agencies. The six areas of focus are:

1. Promoting safe water through CDC's Safe Water System (SWS), which allows individuals, health workers, and schoolteachers to treat and safely store water in homes, health facilities, and schools, and Water Safety Plans (WSPs), which identify water quality threats in

community water systems and water utilities, while implementing solutions to those threats.
2. Improving the efficacy, sustainability, and integration of hygiene and sanitation interventions into communities and institutions, such as schools.
3. Deploying emergency response and outbreak investigation teams at the request of foreign governments and U.N. agencies.
4. Identifying WASH-related factors needed to control or eliminate Neglected Tropical Diseases (NTDs) like Guinea worm disease, trachoma, and intestinal worm infections, which impact hundreds of millions of people around the world.
5. Developing model programs and materials for public health staff training and community health promotion.
6. Investigating the causes of illness, such as diarrhea, to provide critical health data for decision making.

With international cooperation to provide safe drinking water through the development of proper sewage systems and water sanitation systems, the deaths from diarrheal illnesses in developing nations can continue to be reduced globally, with the hope of eradicating further outbreaks and death from cholera.

Chapter IV

The Spanish Flu

Epidemiology

Influenza (colloquially termed the "Flu") is a common viral infection that spreads through a wide variety of mammals including humans, pigs, horses, dogs, seals, and anteaters. Classified as an acute respiratory disease, influenza attacks and infects the respiratory tract, specifically the lungs, of mammals and has the potential to induce pneumonia. Although several influenza outbreaks occurred thousands of years before, the most devastating and important influenza pandemic emerged in February 1918. Decades after the pandemic, researchers utilized genetic analysis of frozen lungs belonging to victims of the Spanish Flu to identify the pandemic-inducing pathogen as the H1N1 influenza virus.

The first major influenza outbreak, and the focus of this chapter, is the 1918 Spanish Flu pandemic. The Flu spread globally and infected an estimated 500 million people (⅓ of the world's population at the time) and caused between 50 and 100 million deaths globally, making it one of the worst pandemics in human history. The Spanish Flu's death rate of 2.5%, compared to 0.1% death rate of other Influenza strains, makes it the most virulent of all pathogenic influenza species (Figure 1).

Death Rates during the Great Influenza Pandemic, 1918–1920

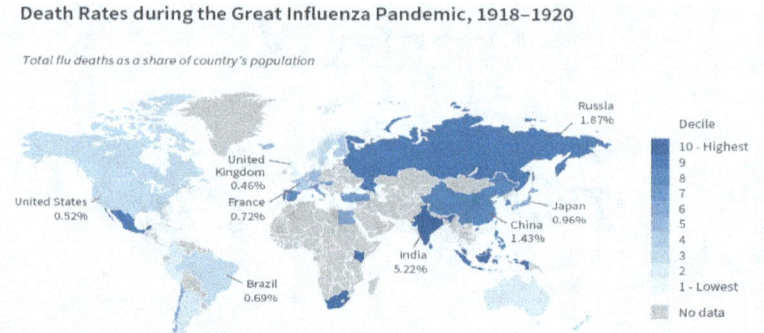

Figure 1. Global death rates of "Spanish Flu". Source: National Bureau of Economic Research

Only three epidemics in world history resulted in mortality approaching or exceeding the mortality caused by the 1918 influenza epidemic: The Plague of Justinian in the sixth century (100 million lives lost over 50 years), the Black Death of 1348-1351 (62 million), and the current AIDS epidemic (25 million). The majority of the victims were young children and the elderly. A distinguishing characteristic of the 1918 epidemic was that it disproportionately killed men and women ages 15 to 44 and led to the biggest drop in life expectancy in the 20th century (Figures 2 & 3).

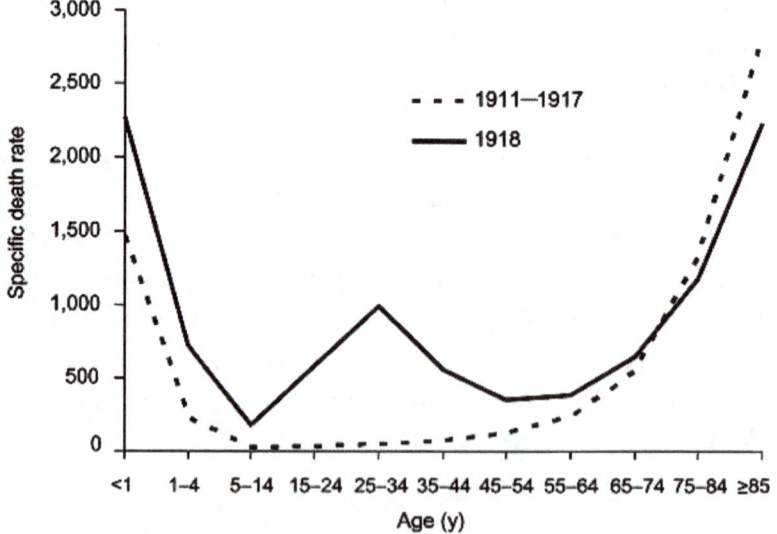

Figure 2. Death rate highest in the 15-44 age groups from Spanish Flu

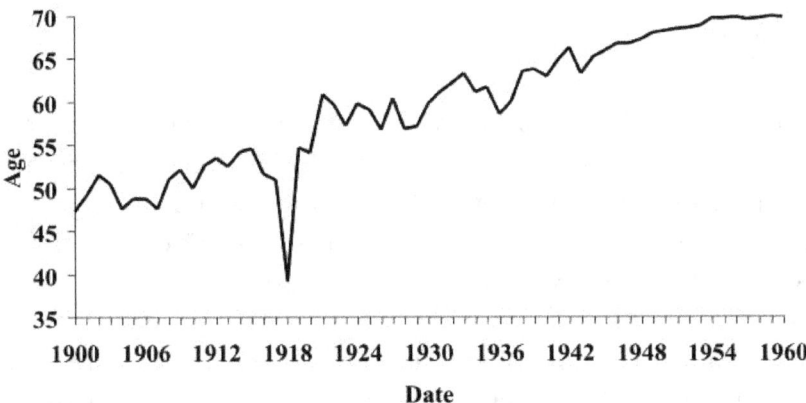

Figure 3. Life expectancy in the United States, 1900-1960 (US Department of Commerce, 1976)

The Hong Kong Flu, the second major Influenza outbreak, surfaced in 1968 and spread to nearly all nations. Analysis of the virus at the molecular level revealed that the virus belonged to the Influenza A family of viruses, the only pandemic-inducing Influenza viruses. The WHO estimates that the Hong Kong Flu killed 1 to 4 million people with a mortality rate of 2%. The virus still circulates to this day as a strain of the seasonal flu, which the common flu vaccine provides immunity against.

In July of 2008, a genetic combination of avian, swine, and human Influenza strains created a genetically novel Influenza strain that would soon be known as the "swine flu". The virus first appeared in humans on March 9th, 2009, in Veracruz, Mexico and the nation's government unsuccessfully attempted to contain the virus. From Mexico, the Swine Flu spread globally, and less than a month later the WHO had declared a global pandemic. Rough estimates state that 7 million to 1.4 billion people had been infected, and of those 284,000 individuals died. The swine flu had a mortality rate of approximately 0.03%.

Origin

When the first cases of the inluenza A (H1N1) pandemic broke out in 1918 during the final year of World War I, the origin of the pandemic was unknown. At that time theories ranged from new form of biological warfare to a by-product of trench warfare resulting from the use of mustard gas.

The Spanish "origin" relates to reports of cases of influenza in the summer of 1918, where as many as eight million Spaniards succumbed to the disease. Even the King of Spain, Alfonso XIII, caught influenza in 1918 (Figure 4).

Figure 4. King Alfonso XIII of Spain (r. 1886-1931)

Unlike most subsequent influenza virus strains that have developed in Asia, the "first wave" or "spring wave" of the 1918 pandemic is thought to have originated in the United States in March 1918. However, the nearly simultaneous appearance of influenza in March to April 1918 in North America, Europe, and Asia makes definitive assignment of a geographic point of origin difficult to discern. On March 4, 1918, company cook Albert Gitchell, possibly patient zero, reported sick with a fever of 104º Fahrenheit at Camp Funston, part of Fort Riley, where 54,000 men were gathered for basic training. Within days, 522 soldiers reported sick and by the end of the month 1,100 soldiers were admitted to hospital with influenza (Figure 5).

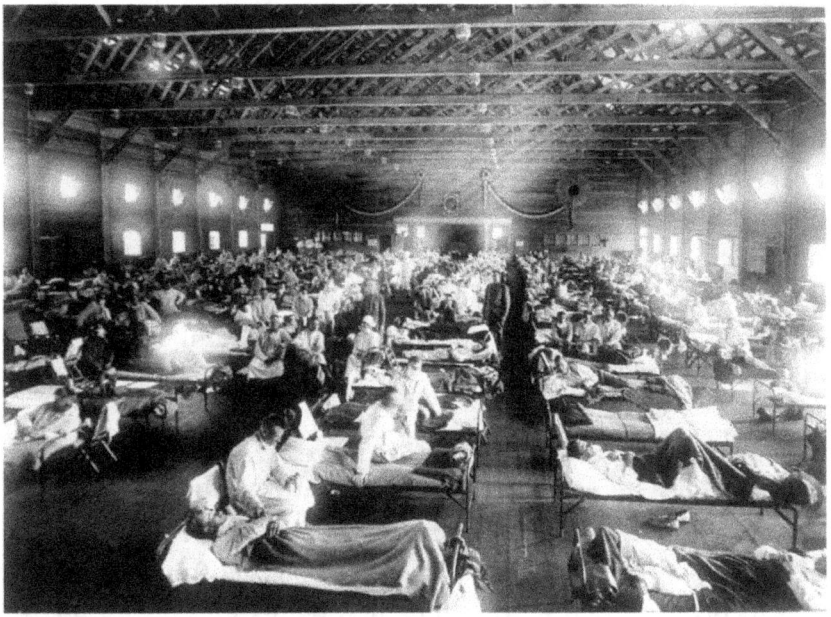

Figure 5. An influenza hospital at Camp Funston.

The wartime context for the pandemic was especially important in the spread of influenza. The flu claimed more lives than the war itself and followed the movement of soldiers around the globe. The American Expeditionary Forces (AEF), which deployed out of Kansas to France, were likely carrying the flu with them in the spring of 1918 as the Allies rushed deployments to halt the German Offensive. The first British cases occurred in mid-April as well, spreading out of ports and Scottish dockyards.

In this 1st phase of the pandemic, termed the "three-day fever", most patients recovered quickly. Their fevers broke after two days, and most were fit for work within a week. Only a minority of patients suffered fatalities from pneumonia caused by the flu. Moreover, by June 1918 the number of cases in Europe and North America began to steadily decline leading to a belief that the flu pandemic was over.

In late August 1918, the 2nd phase of the flu pandemic emerged suddenly across the globe and with much greater lethality. Over the following four months, flu circled the globe infecting approximately 500 million people. Hospitals were overwhelmed. Doctors and nurses disproportionately fell victim to the pandemic while treating an unprecedented number of patients. Pulmonary complications appeared more frequently, contributing to a mortality rate twenty-five times higher than a normal influenza outbreak. Fatalities peaked in November 1918.

Military officials were often the first to realize the severity of the flu, but they did not understand the nature of the illness. The Royal Army Medical Corps began bacteriological examinations of soldiers for "Pfeiffer's bacillus" (*Bacillus influenzae*), which was incorrectly identified as the microbe responsible for the flu in 1892 by the German physician and bacteriologist Richard Pfeiffer.

Virology and Pathology

Influenza viruses are negative strand RNA viruses of the genus *Orthomyxoviridae*. They continually circulate in humans in yearly epidemics, mainly in the winter months. Every few decades, novel virus strains emerge that are highly virulent and lead to global pandemics infection 20 to 40% of the population in a single year. In the United States, influenza is estimated to kill 30,000 people in an average year. Every few years, influenza epidemics boost the annual number of deaths past the average, causing 10–15,000 additional deaths.

The predominant natural reservoir of influenza viruses is thought to be wild waterfowl. Periodically, genetic material from avian virus strains is transferred to virus strains infectious to humans by a process called *reassortment*. Since pigs can be infected with both avian and human virus strains, and various *reassortments* have been isolated from pigs, they have been proposed as an intermediary host in this process. Until recently there was only limited evidence that an avian influenza virus could directly infect humans, but in 1997 eighteen people were infected with avian H5N1 influenza viruses in Hong Kong, and six died of complications after infection.

The Influenza virus' primary method of producing fatalities is by compromising the lungs of the infected. Influenza enters the host's lungs via the nasal cavity or oral cavity. After arriving at the lungs, the Influenza virus enters the host cell and hijacks the machinery of the cell and begins to replicate. Once the Influenza virus reaches its quota of replications, it bursts the host cell and repeats the same process with other infected cells. Healthy immune systems recognize and eliminate the pathogen along with infected tissue. As a result, infected individuals recover from the Flu within several days, unless the immune system's response is too robust and produces too much swelling and fluid accumulation in the lungs, otherwise called edema. Production of edema hinders absorption of oxygen by the lungs because it seeps into the alveoli (air sacs) within lungs that diffuse oxygen into the bloodstream. This results in low oxygen levels in the blood stream (hypoxemia) and eventual death in majority of infected patients (Figure 6).

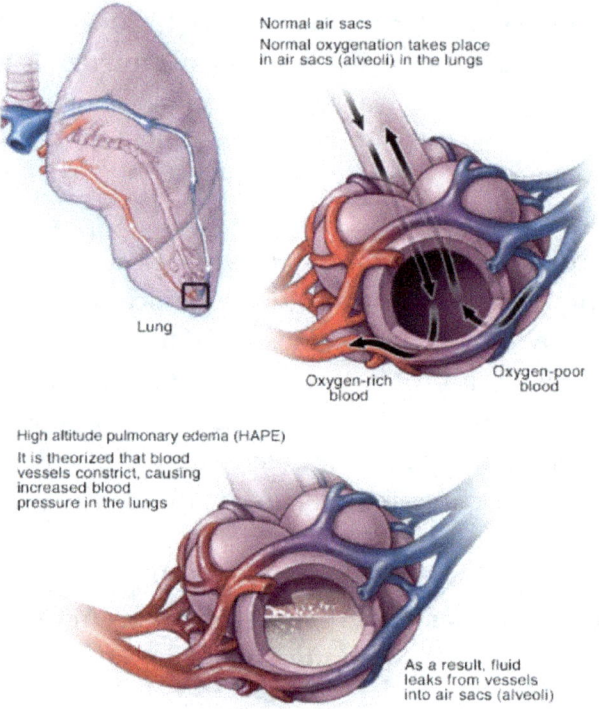

Figure 6. Inflammatory response to H1N1 leading to swelling and fluid buildup in lung tissues resulting in hypoxemia.

Through the reverse genetic engineering of autopsy material belonging to Spanish Flu victims, the molecular structure of the H1N1 pathogen, referred to as the Spanish Flu, was revealed. Analysis of the H1N1 virus identified that the addition of glycosylation sites and genetic mutations in the hemagglutinin glycoprotein separated the H1N1 pathogen from other Influenza viruses. (Figure 7).

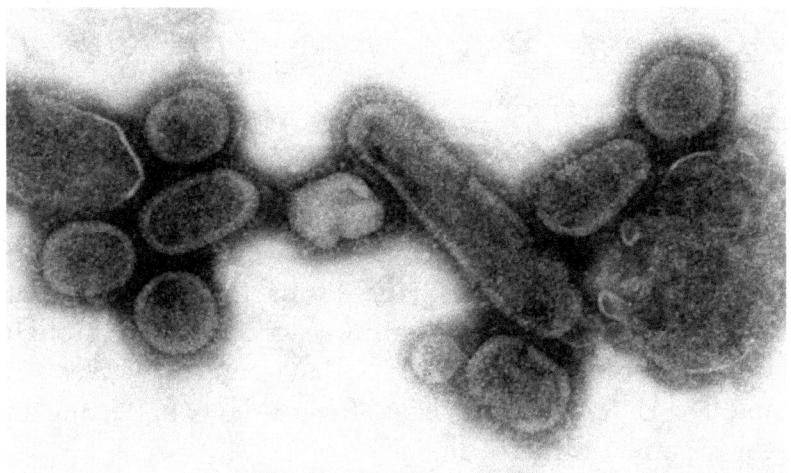

Figure 7. Electron micrograph of H1N1

Influenza viruses rely upon their hemagglutinin glycoproteins to bind to the receptor of epithelial respiratory tract cells and initiate the infection. According to a study done by Taubenberger, the hemagglutinin of the H1N1 pathogen mutated enough to reach a composition that was genetically distinct from other Influenza viruses, which meant the human population was immunologically naïve and had no immune protection. As a result, infected individuals needed to produce new antibodies for the H1N1 pathogen, which delayed the response of the immune system and allowed more time for the virus to replicate in the respiratory tract. The acquisition of the glycosylation site allowed the virus to conceal its antigenic epitope, the part of the virus that the immune system binds to recognize it. This allowed the virus to "hide" from the immune system and continue to attack the respiratory tract.

The majority of individuals who died during the pandemic succumbed to secondary bacterial pneumonia, since no antibiotics were available in 1918. In the hundreds of autopsies performed in 1918, the primary site of damage to the human body from influenza was confined to the lung cells (alveoli), leading to pneumonia and respiratory failure.

Public Health Policy Changes & the Birth of Epidemiology

In the wake of the Spanish Flu, many governments embraced the concept of socialized medicine (healthcare for all, delivered free at the point of delivery). Russia was the first country to put in place a centralized public healthcare system funded through a state-run insurance plan and was later copied by Western Europe. The United States took a different route, preferring an employer-based insurance model.

In 1924, the Soviet government laid out its vision of the physician of the future, who would have "the ability to study the occupational and social conditions which give rise to illness and not only to cure the illness but to suggest ways to prevent it." This vision was gradually adopted across the world, where medical approaches to disease would be not only biological and experimental, but also sociological. This ushered in the concept of public health to develop strategies to avert diseases and gave birth to the scientific field of epidemiology, which has been critical in helping to manage the many pandemics since the Spanish Flu.

The cornerstone of public health is epidemiology—the branch of medicine which deals with the incidence, distribution, and possible control of diseases and other factors relating to health. Epidemiology requires data, and the gathering of health data became more systematic after the Spanish Flu. By 1925, all U.S. states were participating in a national disease reporting system serving as a tool to help recognize early onset of global diseases. Ten years later, reflecting the government authorities' new interest in the population's baseline health, U.S. citizens were subjected to the first national health survey.

Many countries created or revamped their health ministries in the 1920s. This was a direct result of the Spanish Flu outbreak. At that time, public health leaders were not given a voice to shape the strategic response to the pandemic and therefore lacked the ability to sway governments to fund their departments to mobilize a scientific effort to combat the flu. But there was also recognition of the need to coordinate public health at the international level since these contagious diseases didn't respect borders. In 1919, an international bureau for fighting epidemics was formed in Vienna, Austria, the precursor of today's World Health Organization.

The Advent of the Flu Vaccine

At the time of the 1918 pandemic, medical experts did not know its causes, and at the beginning there were no effective treatments. By the end of the pandemic, treatment by transfusion of blood from a survivor was known to be effective. Scientists erroneously thought that influenza was caused by a bacterium they then called *Baccilus influenzae*, now known as *Hemophilus influenzae*. They could isolate and see many types of bacteria under a microscope, but existence of viruses was not known as pathogens since they were too small to be seen by microscope.

In the decades following the pandemic, there was a concerted effort to find the pathogen for the Spanish flu and its treatment. In 1931, virologists Wilson Smith, Christopher Andrews, and Patrick Laidlaw first isolated the Influenza virus. In 1931, American virologist, Richard Shope, had begun studying pigs carrying swine flu. He found that the presence of *H. influenza bacteria* made swine flu worse but did not cause Inluenza as previously speculated. In 1936, Shope showed that all survivors of the Spanish Flu had antibodies to the Influenza virus, thus making the connection that the Influenza virus was the pathogen causing the Spanish Flu.

In 1931, a significant scientific achievement was made at Vanderbilt University. Researchers there found ways to grow the influenza virus in fertile chicken eggs. This meant that they no longer had to get them from sick people or animals. Growing viruses and comparing immunological responses of lab animals, scientists identified two types of flu viruses, naming them A and B. The 1918 pandemic virus was a Type A virus. Today we know that Type A viruses infect both humans and some other animals and are more dangerous, while type B viruses are in humans only.

With the ability to grow large quantities of the viruses, and to identify their characteristics, researchers in the late 1930s began working on a vaccine. In 1938, the US Army began tests of vaccines with a research team that included Jonas Salk. The first mass use of an influenza vaccine for soldiers in the United States came in 1944, and for civilians in 1945 (Figure 8). During the research for this vaccine, it was discovered that immunity against one type of virus does not give immunity against the other. So, the vaccine contained a mixture of both types, as do influenza vaccines of today.

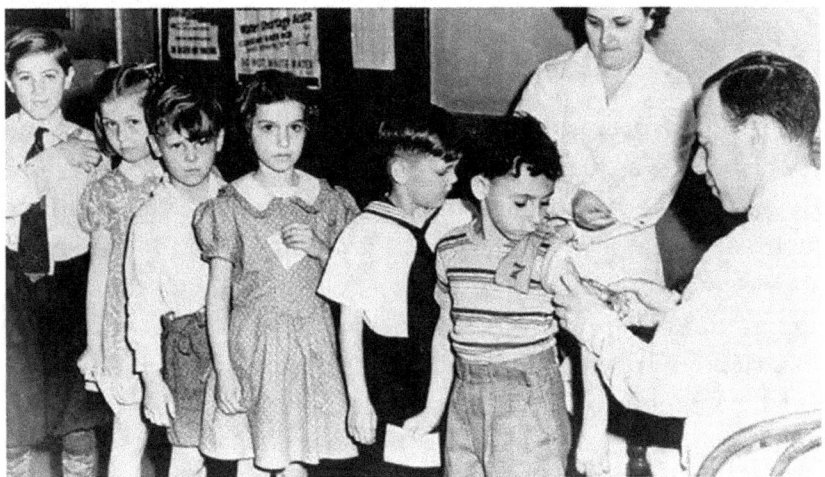

Figure 8. Children in a school in NY City in the late 1940s line up for their flu vaccination. Courtesy of the Library of Congress.

The search for the virus and a vaccine to prevent influenza sparked a number of other discoveries. It created the tools for the development of other vaccines. It also spurred the understanding of the nature of genes and the chemicals that encode them. Oswald Avery, a researcher at the Rockefeller Institute, led the team that discovered in 1944 that deoxyribonucleic acid (DNA) held the genetic code through his research into pneumococcus bacteria. Their research used tools developed in identifying the parts and types of viruses. This deeper understanding of bacteria fostered the development of antibiotics to treat pneumonia. These discoveries spurred on by the hunt to find the cause and to develop a treatment for the Spanish Flu, significantly accelerated scientific knowledge of viruses and bacteria that have served to develop treatments for subsequent pandemics such as HIV, SARS including Covid-19, polio, measles, and smallpox. These scientific developments have had a profound impact on humanity by averting the deaths of millions of people.

Social Distancing & Masking

The primary precautionary measure to limit the spread of the Spanish Flu was to limit social contact. In the United States and globally, governments attempted to close public institutions such as schools, limit large gatherings, and proscribe nonessential meetings. Temporarily closing down schools was not a universally accepted concept at the time, and these "restrictions" placed were lax and for the most part ineffective. Stricter methods of infection control included quarantines and isolation of the ill.

At the individual level, regulating coughing and sneezing along with discouraging sharing utensils and cups were other preventative measures taken to limit the spread of the Flu. The government advised people to disinfect their mouth and nose using chlorine-based disinfectants, a method with no efficacy. Masks at this time, were used, but with little to no understanding of when and where to wear them. Many would cut holes in their masks to smoke cigars (Figure 9).

Figure 9. A man is refused access to a streetcar in 1918 Seattle, because he is not wearing a face mask. Courtesy of NARA.

Globally, civilian public health programs remained conflicted about how to respond to the pandemic. In the United States, the American Public Health Association (APHA) was more proactive in its efforts to prevent the spread of the flu and reduce the severity of the pandemic. The APHA recognized that the disease was extremely communicable and sought to break the spread of infection. It initiated respiratory-hygiene education campaigns about the dangers of coughing, sneezing, and the careless disposal of nasal discharge.

Globally, countries tried a variety of different methods to combat the pandemic. Quarantines, closure of schools, bars, churches, and other gathering places along with compulsory gauze masks were all unsuccessful in containing the disease. High mortality and the concentration of deaths among young adults increased the ability of the pandemic to disrupt the social fabric. The fear of contracting influenza dramatically altered social interactions. Measures of public health, general encouragement from the authorities and the media to avoid inter-personal contacts, and rumors about enemy spies spreading the infection as a kind of biological weapon created a climate of suspicion and mistrust.

The gauze mask was another important preventive tool. The face masks consisted of a half yard of gauze, folded like a triangular bandage covering the mouth, nose, and chin. Some cities, like San Francisco, legislated that everyone should wear a gauze mask in public. A popular rhyme was created to remind people of the city ordinance: *"Obey the laws, And wear the gauze, Protect your jaws, From Septic Paws."*

Figure 10. A gauze mask for protection against influenza worn by a mailman.

In 2007, a study in the *Journal of the American Medical Association* analyzed health data from the U.S. census that experienced the 1918 pandemic and charted the death rates of 43 U.S. cities. That same year, two studies published in the *Proceedings of the National Academy of Sciences* sought to understand how responses influenced the disease's spread in different cities. By comparing fatality rates, timing, and public health interventions, they found death rates were around 50% lower in cities that implemented preventative measures early on, versus those that did so late or not at all. The most effective efforts had simultaneously closed schools, churches, and theaters, and banned public gatherings (Figure 11). This would allow time for vaccine development (though a flu vaccine was not used until the 1940s) and lessened the strain on health care systems.

The studies reached another important conclusion: That relaxing intervention measures too early could cause an otherwise stabilized city to relapse. St. Louis, for example, was so emboldened by its low death rate that the city lifted restrictions on public gatherings less than two months after the outbreak began. A rash of new cases soon followed. Of the cities that kept interventions in place, none experienced a second wave of high death rates (Figure 11).

In 1918, the studies found, the key to flattening the curve was social distancing. And that likely remains true a century later, in the current battle against coronavirus.

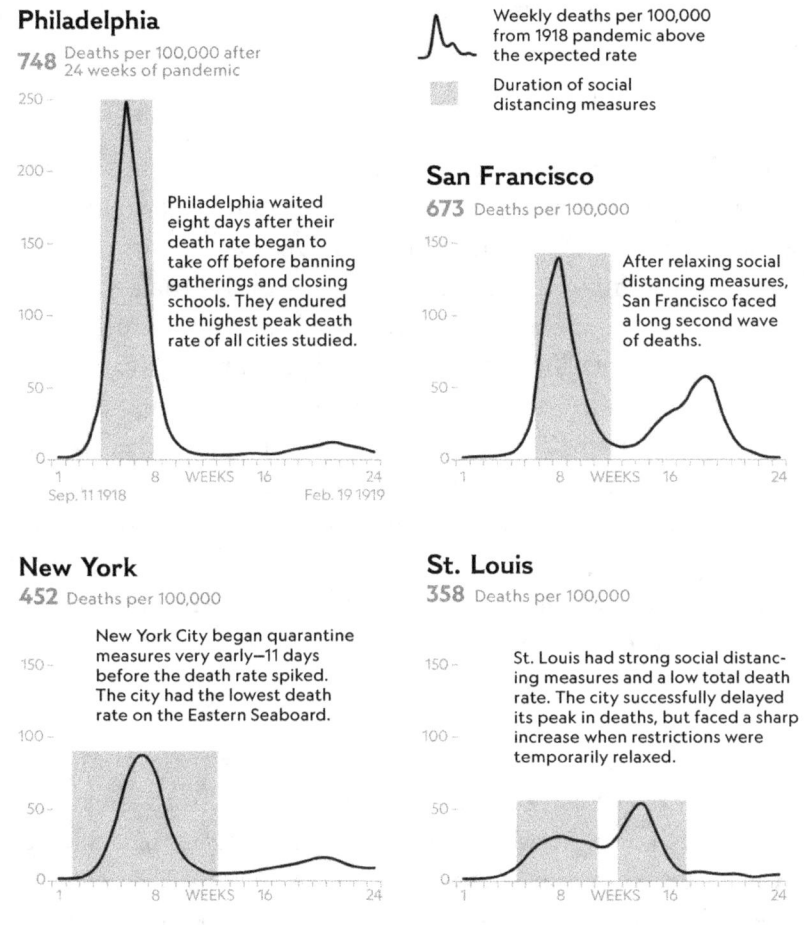

Figure 11. Effects of social distancing in curbing case fatality rates in major U.S. Cities during the Spanish Flu. Source: Champine et. al, JAMA

Economic Implications & the Wealth Gap

Due to its disruptive nature, the Spanish Flu caused a global economic recession and widened the wealth gap between socioeconomic classes throughout all the nations it reached. Severity and longevity of these impacts varied from country to country. Developed countries tended to rebound from economic recessions induced by pandemics relatively quickly, but developing countries experience greater and prolonged economic downfall for decades after the pandemic.

This occurs for a variety of reasons. First, developing countries possess higher population densities and higher rates of poverty. Second, developed countries have infrastructure in place that helps to minimize the contact between individuals in day-to-day life to allow for social distancing, while many undeveloped countries cannot. The combination of higher population density and inability to socially distance resulted in continual recirculation of the virus in underdeveloped countries. This had a profound impact on the economic recovery of developing nations due to loss of labor force and delayed ability to return to work. The higher rates of infection and prolonged duration of the epidemic in these poorer nations suppressed their economic recovery for decades more compared to developed nations.

For economically developed countries at the time, such as the United States, Australia, Britain, and other European nations, the Spanish Flu pandemic did not produce nearly as many economic detriments to the economy. The United States, which suffered a mortality rate of about 0.6%, experienced a 1.5% decline in their GDP along with an approximately 18% decrease in manufacturing output. Furthermore, a country that experienced a much lower mortality rate, such as Australia's 0.3% mortality rate, saw just a 0.66% drop in their GDP. These countries were able to reach their pre-pandemic GDP values within a year. On the contrary, the greater population densities belonging to underdeveloped countries led to greater economic downfalls. For example, India's GDP dropped 12.9% over the span of the pandemic and it did not return to pre-pandemic numbers until 1922, four years after the arrival of the Spanish Flu.

This data set demonstrates mortality and economic distress run parallel to each other, which explains why developed countries rebound from their recessions at a much faster rate and why pandemics produce much more profound impacts in underdeveloped countries

Implications of Trust and Impact on Economic Recovery

Social trust between citizens and between citizens and their governments greatly impacts the ability of nations to recover from pandemics. To quantify the degree of social trust within a nation, the social trust metric was developed. This metric specifically quantifies the degree by which an individual in society trusts another individual. Without trust, individuals question the intentions of others and leads to less interactions between citizens, thereby reducing commercial exchange. As a result, lower social trust values directly correlate with less prosperous economies.

A study completed by Aasve et al, 2021, concluded that for every death per thousand, the Spanish Flu lowered the social trust level by 1.4 to 2.2%. The same study also concluded the Spanish Flu, by causing social disruption and generalized mistrust, had permanent consequences on individual behavior in terms of lower social trust and this lack of social trust constrained economic growth for many decades that followed.

In the aftermath of the Spanish Flu, there was a period of robust economic and social prosperity for many of the nations that were able to quickly rebound from the disease's wrath. The United States entered the Roaring Twenties, Germany entered its Golden Twenties, and Britain experienced a post-war period of prosperity. Interestingly, these nations experienced periods of prosperity *despite* global decline in social trust, high death tolls from World War I and the Spanish Flu, and the contraction of their economies from these two global events. These periods of prosperity can be attributed to quicker containment of the virus, lower death rates, quicker return of the labor force to work and GDP recovery. However, a major driving force that likely resulted in a brisk economic recovery was national pride stemming from trust of the people and the government of these nations. This, in combination with a reviving economy, sparked a general sense of relief and even joy throughout developed countries that sparked these economic booms.

For developing countries that experienced the economic downfall associated with the Spanish Flu for a prolonged amount of time, the social impacts varied but stoked fear rather than trust. In Africa for example, the preventative measures implemented to prevent the spread of the Spanish Flu failed to contain the virus and led to growing mistrust between its citizens' and their governments. Their remained a sense of gloom in these countries, rather than hope. This mistrust has repercussions still seen today in these nation states.

Pandemics have widened the wealth gap between the developed and developing nations. Historically, all pandemics have served as stress tests for all nations. They particularly inflame the ongoing social and economic struggles within developing nations.

Chapter V
Human Immunodeficiency Virus

AIDS was first recognized in the summer of 1981. Young gay men began falling ill and dying of opportunistic infections more commonly seen in people with weakened immune systems. They became emaciated and often developed dark purple lesions on their arms and faces due to an aggressive form of cancer called Kaposi's sarcoma (Figure 1). Physicians were baffled and fear spread through the gay community with every new case of what was termed the "gay plague." Doctors treated one opportunistic infection after another, but these patients rapidly deteriorated and eventually succumbed to their infections.

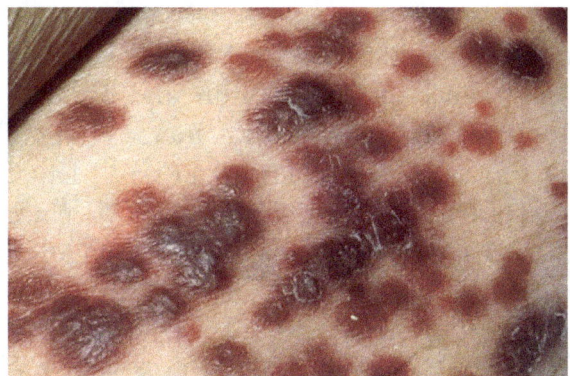

Figure 1. Kaposi's Sarcoma

By late 1982, epidemiologic evidence indicated that AIDS was an infectious disease transferred by bodily fluids and by exposure to contaminated blood or blood products. Without a test for AIDS, blood banks had difficulty safeguarding the blood supply. Thus, the blood supply remained unsafe for years, and many people were transfused with contaminated blood. One such person was Elizabeth Glaser, who in turn transmitted the disease to her unborn child. Their deaths triggered an outcry for more research and ultimately led to the formation of the Elizabeth Glaser Pediatric AIDS Foundation. Approximately 15,000 hemophiliacs in the US became infected with HIV because of transfusions with contaminated blood products between 1981 and 1984. Similarly, intravenous drug users were at high risk due to exchange of blood during needle sharing.

In 1983, the cause of AIDS was discovered. Using tissue from a "pre-AIDS" patient with acute lymphadenopathy, Luc Montagnier and colleagues at the Pasteur Institute detected a new human retrovirus. In May 1986, the International Committee of the Taxonomy of Viruses recommended that this virus be named the human immunodeficiency virus (HIV). Isolation of HIV enabled the development of the first blood test in 1985 to identify infected persons and contaminated blood supplies.

In 1996, antiretroviral drug regimens consisting of various combinations of reverse transcriptase inhibitors and protease inhibitors referred to as active antiretroviral therapy (ART) were developed. ART transformed AIDS from an acute, lethal disease to a chronic, manageable infectious process. Vaccines have been developed and have been 25% effective in preventing HIV infections, but cure remains elusive.

Origins

In 1999, researchers found a strain of Simian Immunodeficiency Virus in a chimpanzee (called SIVcpz) that was almost identical to HIV in humans. The researchers who discovered this connection concluded chimpanzees were the source of HIV-1, and that the virus had at some point crossed species from chimps to humans. The same scientists then conducted more research into how SIV could have developed in the chimps. They discovered that the chimps had hunted and eaten two smaller species of monkeys, red-capped mangabeys and greater spot-nosed monkeys. These smaller monkeys infected the chimps with two different strains of SIV. The two different SIV strains then joined together to form a third virus (SIVcpz) that could be passed on to other chimps. This was the strain that was determined to infect humans.

The most accepted theory of how HIV crossed from chimps to humans is the 'hunter' theory. SIVcpz is believed to be transferred to humans from chimps being killed and eaten, or their blood getting into cuts or wounds on people while hunting. It is theorized that the virus mutated and adapted itself within the human host and became HIV-1.

The most studied strain of HIV is HIV-1, which is the strain that has spread throughout the world and is responsible for the vast majority of HIV infections today. HIV-2 comes from SIVsmm in sooty mangabey monkeys rather than chimpanzees. The crossover to humans is believed to have happened through the butchering and consumption of monkey meat. It is far rarer, and less infectious than HIV-1. As a result, it infects far fewer people, and is mainly found in a few countries in West Africa like Mali, Mauritania, Nigeria and Sierra Leone.

Studies of some of the earliest known samples of HIV provide clues about when it first appeared in humans and how it evolved. The first verified case of HIV is from a blood sample taken in 1959 from a man living in what is now Kinshasa in the Democratic Republic of Congo. The sample was retrospectively analyzed, and HIV was detected. There were numerous earlier cases where patterns of deaths from common opportunistic infections suggested that HIV was the causative agent.

Using the earliest known sample of HIV, scientists have been able to create a detailed map of HIV transmission, allowing them to discover where HIV started. Their studies concluded that the first transmission of SIV to HIV in humans took place around 1920 in Kinshasa in the Democratic Republic of Congo. The same area is known for having the most genetic diversity in HIV strains in the world, reflecting the number of different times SIV was passed to humans.

The area around Kinshasa is full of transport links, such as roads, railways and rivers. The area also had a growing sex trade around the time that HIV began to spread. The high population of migrants and sex trade might explain how HIV spread along these infrastructure routes. By 1980, half of all infections in Democratic Republic of Congo were in locations outside of the Kinshasa area.

In the 1960s, HIV-1 made its way to Haiti. At this time, many Haitian professionals who were working in the colonial Democratic Republic of Congo during the 1960s returned to Haiti. Initially, they were blamed for being responsible for the HIV epidemic, and suffered severe racism, stigma and discrimination as a result. HIV-1 is now the most geographically spread subtype of HIV internationally. By 2014, this subtype had caused 75 million infections.

Virology

HIV belongs to a class of viruses known as retroviruses and a subgroup of retroviruses known as lentiviruses or "slow" viruses. HIV is an enveloped, single-stranded positive-sense RNA virus (Figure 2) with a genome of 9749 nucleotides in length that encodes a total of nine viral proteins. The HIV genome contains three major genes including gag, pol, and env, encoding major structural proteins as well as essential enzymes. The gag gene encodes viral core proteins, the pol gene encodes a set of enzymes required for viral replication, and the env gene encodes the viral surface glycoprotein gp160.

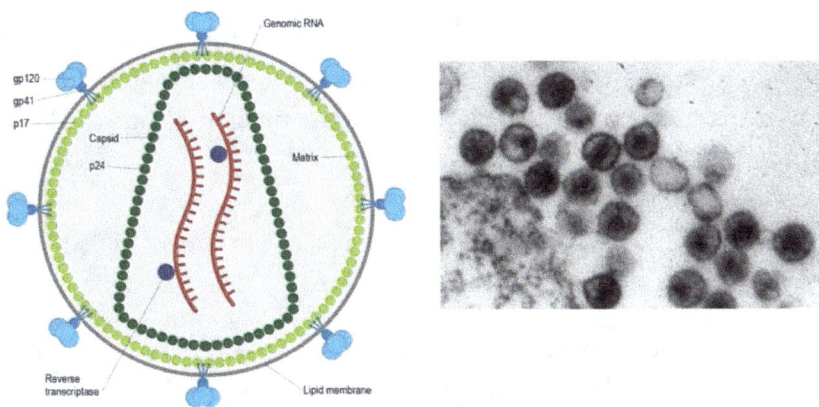

Figure 2: HIV-1 illustration and electron-micrograph of HIV particles.

The course of infection with these viruses is typically characterized by a long period between initial infection and the onset of serious symptoms. Like all viruses, HIV can reproduce only inside cells by hijacking the cell's machinery. The first step of infection is the binding of gp120 to the CD4 receptor of the cell, which is followed by the fusion of the virus and cell membrane and is mediated by the gp41 molecule. The virus then penetrates the cell and begins uncoating. Next, reverse transcription takes place to convert their viral RNA into DNA. Viral DNA is then incorporated into the host cell genome, and provirus synthesis begins. This is followed by the synthesis and maturation of virus progeny (Figure 3).

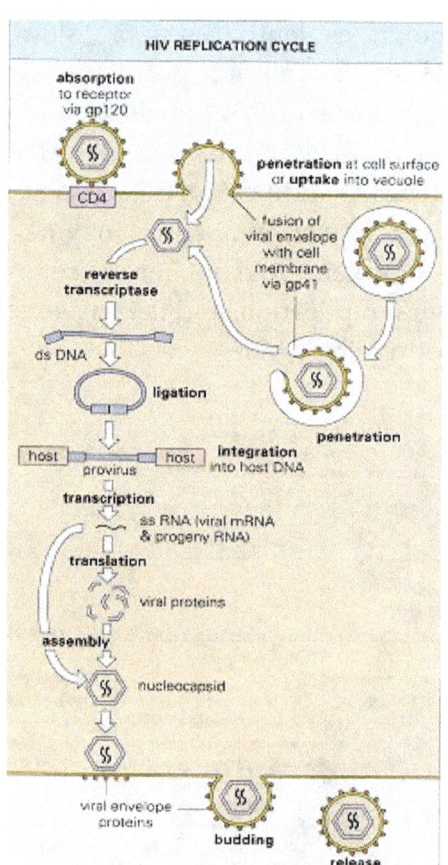

Figure 3. Source: Mims C, Playfair J, Roitt I, Wakelin D, Williams R. 1993. Medical Microbiology. p. 24.16

The profound immunosuppression seen in AIDS is due to the depletion of T4 helper lymphocytes (CD4 cells). CD4 cells are an essential part of the immune system that directs the immune cells to attack and destroy viruses, bacteria and fungi. In the immediate period following exposure, HIV is present at a high level in the blood. It then settles down to a certain low level during the incubation period. During the incubation period, there is a massive turnover of CD4 cells, whereby CD4 cells killed by HIV are replaced efficiently. Eventually, AIDS develops when killed CD4 cells can no longer be regenerated and as the CD4 cell count drops, the infected individual cannot fight off infections and succumbs to bacterial, viral and fungal injections.

Transmission of HIV requires contact with a body fluid that contains either infectious virus or HIV-infected cells or a combination of both. HIV can appear in nearly any body fluid, but transmission occurs predominantly through blood, semen, vaginal and rectal fluids, and breast milk. No case of HIV transmission has been traced to the coughing or sneezing of an infected person or to a mosquito bite. The three main routes of HIV transmission are parenteral exposure (e.g., blood transfusion, needle sharing), unprotected sexual contact, and mother to child transmission. Sexual exposure is the most common route of infection and drives the HIV pandemic in most countries, followed by needle sharing IV drug use.

When a person becomes infected with HIV there is an initial burst of viral replication, and the individual may experience flu-like symptoms for a few weeks. Antibodies to the virus are detectable from between 6 and 18 weeks after the initial infection. There is also a slow and steady decline in the level of CD4+ cells. As mentioned above, CD4+ cells are critical component of the human immune system. These cells secrete chemicals that help to stimulate the immune response and coordinate the immune system to attack and destroy pathogens. Therefore, a reduction in CD4+ cells translate into a reduced immune response to pathogens, leading to an impaired immune system response to infections and cancers.

The onset of HIV antibody production is followed by a long but variable incubation period of about 10 years. This latency period is followed by an early symptomatic stage in which there is a renewed decline in CD4+ cells and an increase in infections, which are generally not life-threatening. A late symptomatic stage of HIV disease follows that is characterized by more serious infections and a continuing decline in CD4+ cells. What is often referred to as "full-blown AIDS" is the end stage of a progressive and continuous process of profound immune deficiency. At this stage, CD4+ cell counts are very low, and opportunistic disease and infection pose the greatest threat to the patient's survival.

Initially, HIV infection was considered fatal within 10-15 years. However, most industrialized countries, where patients have treatments available to them, have seen a deceleration in the progression to full-blown AIDS and a decline in the death rate over the past decade.

Epidemiology

Approximately 76 million people worldwide have been infected with HIV since the start of the pandemic in 1981, and 38 million living with HIV today. An estimated 35 million people died from AIDS related infections since the start of the pandemic. Although there has been a significant decline in new infections since the mid-1990s, there were still about 1.7 million new infections in 2019, or about 5,000 new infections per day (Figure 4). Recent data show that while progress has been made, that progress is unequal within and between countries. Over the past 20 years, the global response has vastly slowed the spread of HIV, averting more than 30 million new infections and nearly 8 million deaths (Figure 5). The HIV/AIDS pandemic has affected sub-Saharan Africa most substantially, where it appears to be stabilizing. HIV remains a leading cause of death worldwide and the leading cause of death globally among women of reproductive age. However, AIDS-related deaths have declined, due in part to antiretroviral treatment (ART) scale-up. In 2019, 690,000 people died of AIDS, a 37% decrease from 1.1 million in 2010 and a 59% decrease from the peak of 1.7 million in 2004.

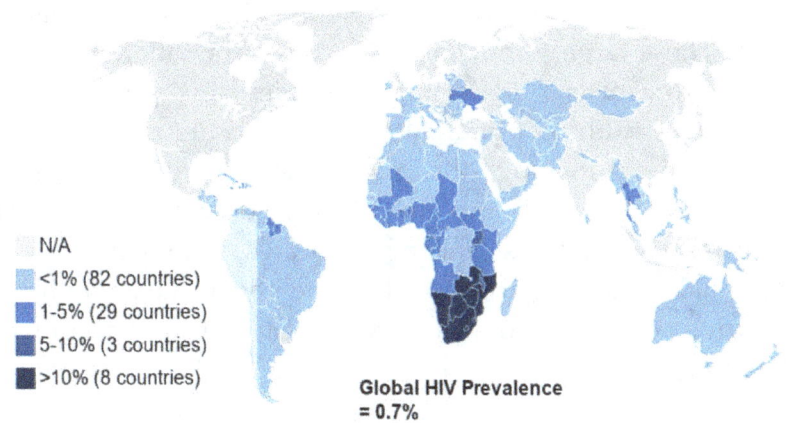

Figure 4. Prevalence of HIV Globally, 2019. Source: Kaiser Family Foundation

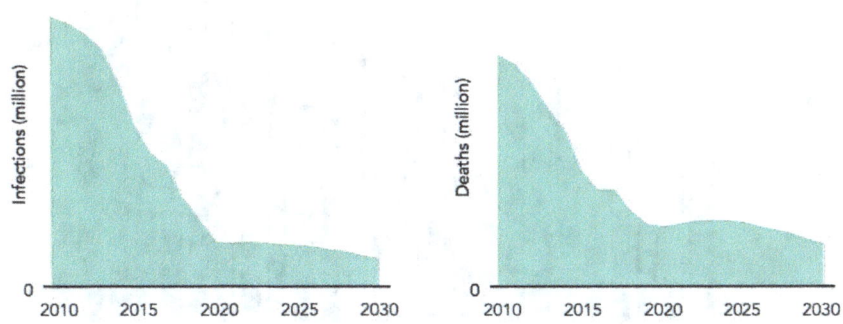

Figure 5. Decline in HIV infections and deaths since 2010 and projected to 2030.

Since the discovery of AIDS, gay and bisexual men have been disproportionately affected. In the United States, the Kaiser Family foundation states that while gay men make up 2% of the population, they account for 66% of new HIV infections, 56% living with HIV, and 55% of all AIDS deaths since the beginning of the epidemic. This is likely due to higher transmission and blood exchange associated with anal intercourse.

It is estimated that, in the next 20 years, about 70 million people will die from AIDS, of whom 55 million will be in sub-Saharan Africa. These large numbers will cause a dramatic decline in average lifespans, eradicating decades of progress in population health that had led to improved life expectancies. Barriers to containing the pandemic include the large proportion of asymptomatic individuals serving as vector for HIV transmission, the social stigma of HIV, criminalization of individuals testing positive for HIV, and governmental policies restricting access to testing, access to health care and treatment (Figure 6).

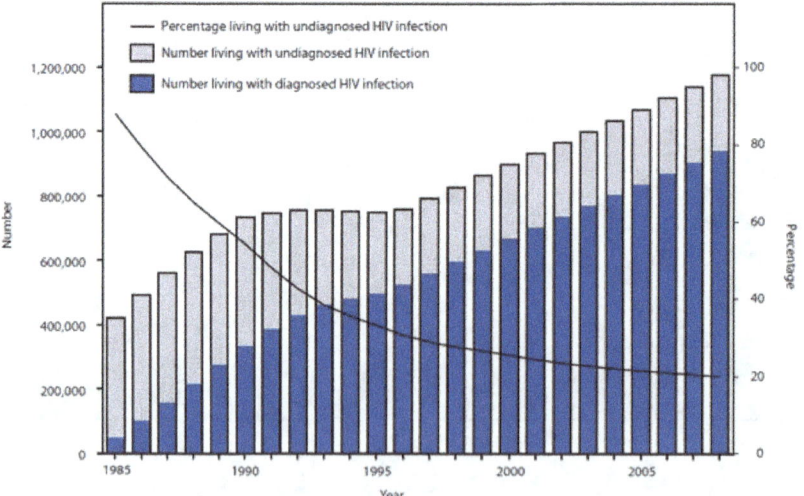

Figure 6. Number of undiagnosed cases since discovery of HIV

Therapeutics

In the 1960s, a failed anti-cancer drug called zidovudine was shelved and rejected by the medical community. Twenty years later, under the name AZT, it became the first of many medicines in the fight against HIV. AZT was seen as a breakthrough as the first medication that seemed to attack HIV replication. It was the first real hope for people infected with HIV, which until then had almost certainly led to death. AZT had a staggering cost at $8,000 per year and is said to be the most expensive prescription drug in history. In 1986, a clinical trial on patients in eight American cities was stopped after four months because AZT seemed to have such dramatic effects on the virus. The next year AZT became available commercially as the first antiretroviral to be registered by the US Food and Drug Administration as Retrovir. Unfortunately, follow-up research was not so optimistic. Clinical trials in Europe found no long-term benefit from using AZT, especially if patients started taking the drug before they showed signs of AIDS. In fact, healthy patients taking AZT seemed to die faster than their sicker counterparts. What was not known at that time was that taking only one antiretroviral at a time has a short-term benefit, because the virus mutates so rapidly that drug-resistant strains swiftly become dominant.

Until 1991, AZT was the only hope available for people sick with HIV and dying from AIDS. In that same year, another antiretroviral, (Didanosine, Videx), was created specifically for patients who had become resistant to AZT. In 1992, ddC (zalcitabine, Hivid), was approved for use in the US, followed by d4T (Zerit) in 1994 and 3TC (Epivir, lamivudine) in 1995. These drugs are classified as nucleoside reverse transcriptase inhibitors (NRTIs) that resemble the chemical building blocks of DNA and RNA (nucleosides) used by reverse transcriptase, a key enzyme required by HIV for intracellular replication.

NRTIs have a chemical twist, which ensures that once taken up by the enzyme, the NRTI molecules terminate the building of the viral DNA chain, stalling virus production.

The problem is that HIV has the ability to quickly mutate and develop drug resistance to different drugs in the same group. For example, a patient treated with AZT alone is likely to quickly show some resistance to other NRTIs. As they are all slightly different, dual therapy using two drugs in the same class is still more effective and sustainable than monotherapy. This was demonstrated in 1997, when the FDA registered Combivir, a combination drug containing both AZT and 3TC. In this case, resistance to one drug appeared to counter resistance to the other.

In 1996, the first in a new class of antiretrovirals called non-nucleoside transcriptase inhibitors (NNRTI) was developed and brought to market. NNRTIs are a group of drugs that stop the duplication of viral DNA by directly disabling the reverse transcriptase enzyme itself. This was an extremely important breakthrough, as the virus had become resistant to NRTIs. Researchers quickly found that dual therapy (using two drugs simultaneously) was most effective when the drugs were from two different groups.

HIV infection has developed from a fatal into a manageable chronic disease with life expectancy, in some instances, estimated to be near that of the general population. Treatment with ART is life long as it only suppresses the replication of HIV but does not eradicate or cure the infection. Stopping ART results in HIV viral load rebound, progressive $CD4^+$ T-cell count decline, and clinical disease progression. One of the primary aims of ART treatment is to maintain health by preventing clinical disease progression at low cost of drug toxicity. This is achieved by inhibiting viral replication by antivirals, resulting in long-term suppression of plasma viral load. Treatment success is defined as maintaining plasma viral load at an undetectable level (<50 copies/mL) and by reconstitution of the immune system.

Although ART inhibits HIV replication and prevents disease progression, it does not eliminate the virus completely from infected patients, predominantly because of the presence of latently infected resting memory $CD4^+$ T cells. These infected cells contain HIV viral DNA within their chromosomes but usually express little or no viral RNA and no viral proteins. However, if ART is discontinued, these cells are stimulated to produce infectious virus and kill CD4 cells leading to progressive immune suppression and recurrence of opportunistic infections.

The world's first long-acting monthly injectable HIV treatment regimen received FDA approval in June 2021. Pivotal Phase 3 studies showed that among more than 1,100 patients, the injectable regimen—consisting of two antiretroviral drugs, one developed by Janssen and the other by ViiV Healthcare—was as effective in suppressing viral replication as a daily three-drug oral regimen. The approval represents another advance in the fight against HIV and a new way to lessen the burden of those living with the virus.

Currently, efforts to find a cure for HIV have not borne fruit. A study published in Lancet in March 2021 discovered that 4% of the population in the Democratic Republic of Congo is innately immune to HIV. These individuals tested positive for HIV antibodies but had undetectable viral counts and were never treated with antiretroviral drugs. This holds some promise in discovering a possible cure if the mechanism of immunity in this population can be established.

Biomedical Advances

In addition to the advances made in the HIV therapeutics, the years of HIV research has translated into advances in oncology and other immune diseases. Some of the medical advances include the following:

Gene Therapy

Inactivated HIV virus has been used for life saving gene therapy by inserting "healthy" genes into individuals with mutated genes. Cells are removed from the patient and cultured with the inactivated HIV with the functional gene inserted into the virus (ex-vivo) and then re-inserted into the patient (in-vivo) (Figure 7).

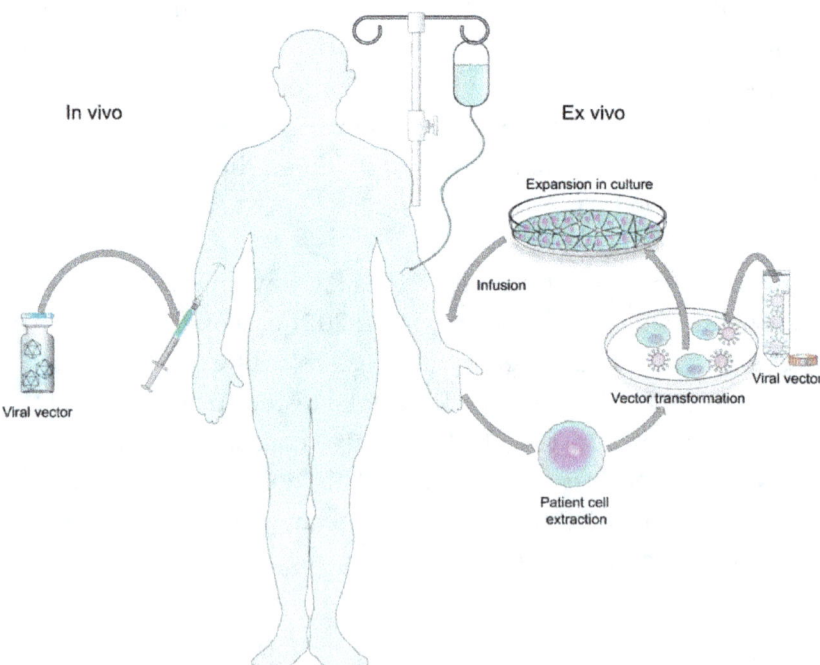

Figure 7: Gene therapy methodology. Bulcha et. al, 2021

On such example is the gene therapy treatment of Severe Combined Immunodeficiency due to Adenosine Deaminase Deficiency (ADA-SCID). This is a fatal disease in young children that cannot mount an immune response to common bacteria and viruses resulting in premature death. Kohn et al. (NEJM, 2021) used inactivated HIV as a vector to insert a copy of a functional adenosine deaminase gene to replace the mutated one that these children were born with. Overall survival was 100% in those children that successfully incorporated the gene. Without this treatment, 100% of these children would have died within their first 5 years of life.

There have been case reports of patients being cured of cancer using inactivated HIV gene therapy. In 2012, a young girl with acute leukemia wasn't responding to conventional treatments and was on the brink of death. She was given an experimental treatment using an anti-cancer gene that could only be delivered to her cells using disabled HIV as a carrier. Within weeks she was cancer-free. In 2014, a Utah man with the same type of aggressive leukemia received a similar treatment using disabled HIV also went into remission from his cancer. Additionally, six HIV-negative children with usually fatal genetic immune disorders (Wiskott-Aldrich syndrome and chronic granulomatous disease) were cured using similar targeted gene therapies, all of which employed inactivated HIV. HIV gene therapy vectors that use inactivated HIV to alter T lymphocytes in order to invoke an immune response are now FDA approved for the treatment of certain cancers, such as acute lymphoblastic leukemia

Treatment & Cure of Viral Diseases

HIV/AIDS therapies are critical in the treatment of other diseases. For example, three drugs developed to treat HIV—lamivudine, tenofovir, and entecavir—are now the mainstays of therapy for hepatitis B virus (HBV) infections. Another antiviral drug called adefovir, which failed as an HIV treatment, was found to suppress HBV at much lower dosages and has been approved for the treatment of chronic HBV disease. More recently, a drug called sofosbuvir that is modeled on HIV reverse transcriptase inhibitors, one of the main classes of anti-HIV drugs, is being used to treat and cure hepatitis C.

Development of PCR

One particle of HIV genetic material can be located among millions of other particles by using extraordinarily sensitive techniques known as PCR (polymerase chain reaction) and RT-PCR (reverse transcription PCR). PCR tests, initially developed for diagnosing HIV, are now routinely used to rapidly detect several other infectious diseases, including COVID-19, hepatitis C, tuberculosis, chlamydia, influenza, Lyme disease, and many fungal infections.

These techniques have also made it possible to measure otherwise undetectable levels of cancer cells in people who appear to have been cured. This detection allows doctors to initiate new therapy or to continue ongoing treatments that might otherwise have been discontinued.

Furthermore, the discovery of HHV-8, a herpes virus linked to Kaposi's sarcoma, was made possible by a new application of PCR. This technique is now being used worldwide to seek possible infectious causes for diseases of unknown origins.

In summary, AIDS research is providing insights into a range of diseases, their causes, and their treatment. Better diagnostic methods, therapies to restore the immune system, newer preventive antibiotics and drugs, and new treatments for infectious diseases and cancer, all developed in the course of AIDS research, are improving and prolonging countless lives every day.

Criminalization of a Virus

HIV has been and still is considered a disease of the undesirables and has led to the criminalization of gay communities and those infected with HIV globally. Unlike the HIV pandemic, pandemics arising from other viruses infect people of all socio-economic status, gender, and sexual preference due to their ability to easily spread from respiratory droplets or contact with infected surfaces or clothing. HIV, on the on the other hand, is much harder to transmit as it requires exchange of bodily fluids and blood contaminated with HIV viral particles. Since it was first reported in the gay community, HIV was given the stigma of being a byproduct of the gay lifestyle. Despite data determining that HIV was and is predominantly transmitted through heterosexual behavior with the highest case counts in Sub-Saharan Africa, the stigma persists as a disease of gay men and of the poor. This has thwarted efforts to contain the disease due to fears of reprisals of infected individuals being ostracized and possibly being imprisoned in certain communities and nations. Unlike other pandemics that have catalyzed changes for the better in the global communities, HIV has exposed the prejudices within societies and nations that have catalyzed changes for the worse and have led to social regression and propagation of the disease.

HIV exposure and transmission is criminalized in 68 countries. There are travel bans for infected individuals in 20 countries and mandatory HIV testing for individuals applying for marriage, work or resident permits in various other countries. The premise of these laws is to avert perceived unacceptable behavior through fear of punishment. However, studies have shown that HIV criminalization undermines the response to HIV treatment and containment. (UNAIDS, 2019).

In 2018, 67 countries criminalized same-sex behavior and eight countries imposed the death penalty. In parts of the Caribbean, homosexuality is criminalized, and it is estimated 25% of homosexual men are infected with HIV, which is significantly higher than countries that do not criminalize same sex conduct. In 2016, Indonesia imposed a systematic crackdown on the LGBT community with raids on gay bars and saunas. This has made it more difficult for these communities to receive basic health care and no HIV care.

The impact of the stigma of having HIV has not only led to impaired access to healthcare, but also to basic human rights. In some countries, people with HIV are refused the right to work. As a result, in some regions, a large proportion of people living with HIV are unemployed. The lack of employment has also hampered the ability of those infected with HIV to afford to get treatment with antiretroviral drugs and is most likely promoting the further spread to the disease. In some countries, women are prosecuted for transmitting HIV to their baby despite having no access to HIV care that would have prevented transmission with antiretroviral therapy (UNAIDS, 2018).

As a result of these social barriers, the percentage of people with confirmed HIV that are receiving treatment with ART is significantly reduced (Figure 8). Till these social barriers are eliminated, the route to eradicating HIV will be a difficult one, if not impossible. If the model of mass contact tracing and treatment (vs. vaccination) as done with smallpox is followed in tackling the HIV pandemic, then there is the definite potential to eliminate HIV. By reducing viral load to undetectable levels with ART, the ability to transmit HIV from person to person and form mother to child will diminish to less than 1%. Unless there is complete access to testing and antiretroviral medicines to suppress viral load and thereby transmissibility, eradication of HIV will remain elusive.

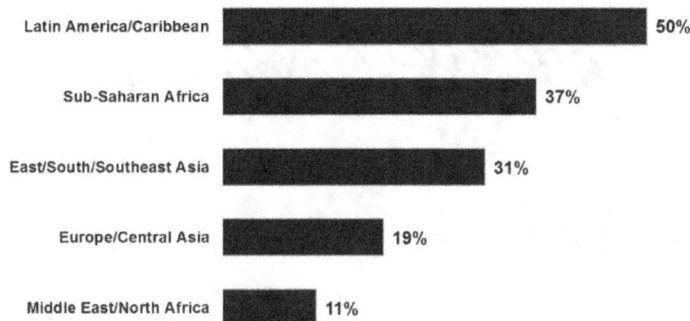

Figure 8. Percentage of individuals testing positive for HIV receiving Anti-retroviral therapy (ART) per region.

Chapter VI

SARS & Covid-19

Epidemiology

Coronaviruses are a class of genetically diverse viruses found in a wide range of host species, including birds and mammals such as pigs, camels, bats and cats. Many coronaviruses cause intestinal and respiratory infections in animals and humans. The first coronavirus (CoV) pandemic appeared in 2002 - 2003 when a cluster of cases of pneumonia were first reported in the Guangdong Province of China and later in Hong Kong. Researchers in Hong Kong isolated a novel coronavirus, and the disease was later renamed severe acute respiratory syndrome (SARS-CoV) due to its ability to infect lung cells resulting in acute respiratory failure. Within 11 weeks from the first SARS case in Hong Kong, it had spread to an additional 27 countries. There were a total of 8,096 probable cases and 774 deaths with a death rate of approximately 1 in 10 infected patients (10%). Sixty-six per cent of the cases occurred in China, 22% in Hong Kong, 4% in Taiwan and 3% in both Singapore and Canada. Chinese scientists traced the virus through its intermediary host the Asian palm civets to the primary host the cave-dwelling horseshoe bats in Xiyang Yi Ethnic Township in the Yunnan Province of China.

A second wave of SARS was detected in Saudi Arabia in September 2012. It spread to 27 countries. From its emergence through January 2020, WHO has confirmed 2,519 cases and 866 deaths with a death rate of about 33%. Molecular and genetic analysis of the virus confirmed it to be a coronavirus and was later named Middle Eastern respiratory syndrome (MERS-CoV). Among all reported cases in humans, 80% occurred in Saudi Arabia. Only 2 people from the United States have tested positive for MERS-CoV. Research evidence suggests that MERS originated in bats and then transmitted to humans through camels (Figure 1).

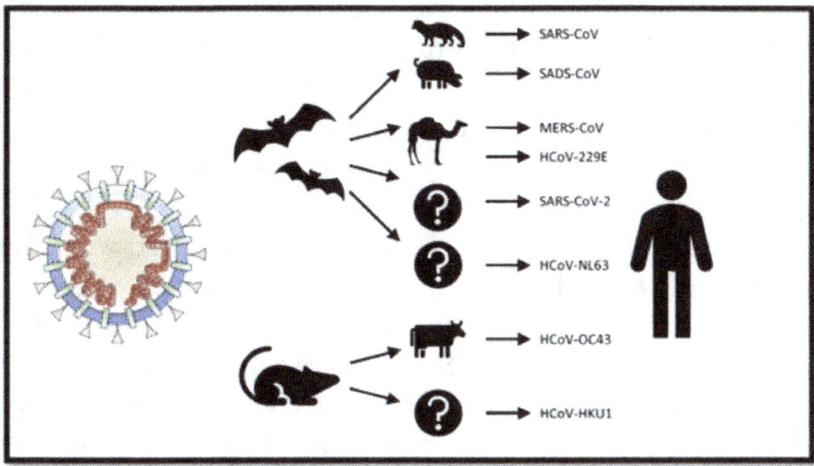

Figure 1. Zoonotic transmission of coronaviruses.

In December 2019, a cluster of cases of pneumonia appeared in Wuhan City in the Hubei Province of China. Virus particles from infected patients were isolated and molecular analysis showed that the pathogen was a novel coronavirus. First named 2019-nCoV and renamed by the WHO as Coronavirus disease 2019 (COVID-19). This novel coronavirus is now the 7th member of the coronavirus family known to infect humans. COVID-19 has been reported in all nation states and has infected 179 million people and killed approximately 3.9 million with a death rate of 2.2%.

Virology & Pathophysiology

Interestingly, the death rate from COVID-19 infection is far less than that of the Hong Kong Strain and MERS at 2.2 % vs. 25% and 33% respectively. However, the total number of infections and deaths associated with COVID-19 are significantly higher compared to Hong Kong CoV and MERS due to its highly infectious nature. Structurally, these viruses are greater than 94% homologous and all contain on their viral capsid (outer coat) Spike (S) proteins (Figure 2). S protein is essential for viral attachment and fusion with the host cell membrane, and subsequent entry of the virus into human lung cells (alveoli). Without this protein, coronaviruses are not infectious. It appears that the extreme contagious nature of COVID-19 is linked to the Spike (S) protein. From the published study by Wabo Tai, et al. (Cellular & Molecular Immunology, March 19, 2020), they identified several point mutations that significantly increase the affinity of the S protein to bind the alveolar cell (lung cell) surface receptors called ACE2 and entry (endocytosis) into the cells. This increased affinity of COVID-19 viral particles to infect lung cells compared to other variants of coronaviruses is the likely cause of the significantly higher infection rates on exposure to COVID-19. Although the percentage of patients dying from COVID-19 infection is significantly less than other coronavirus strains, this high transmissibility (infective rate) has resulted in more deaths by many factors compared to MERS and the Hong Kong pandemics.

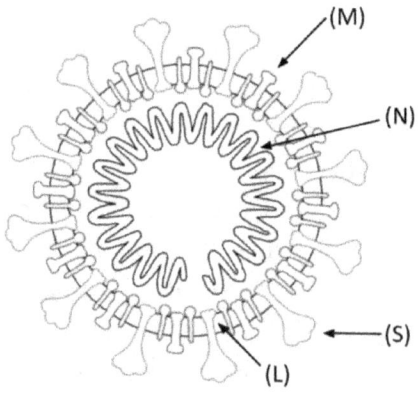

Figure. 2. Viral structure of COVID-19. (M) Membrane protein. (N) Capsid + RNA. (S) Spike protein. (L) Lipid bilayer.

Once lung cells are infected, the virus takes over the machinery of the cell to produce more viral particles. These viral particles kill the cells they replicate in and release newly formed virus that infect neighboring cells. This process continues until the immune system is able to recognize the infected cells and attempts to kill these cells before the virus has time to reproduce and spread the infection. However, it takes the immune system days to possibly 2 weeks to mount an adequate immune response. During the time the immune system is developing neutralizing antibodies, the virus rapidly spreads to infect millions of cells, particularly in the lungs. Once the immune system develops the neutralizing antibodies against the Spike protein, the antibodies latch onto the S protein of infected cell surface and target these cells to be attacked by the immune system.

The lungs are very delicate organs that allow mammals to take oxygen from the air and allow it to travel into red blood cells. However, in order for oxygen to travel from the lung cells into red blood cells, there needs to be very little distance between the air in the lungs and the arteries of the lungs. As a result, the distance from the lung to the arteries is only two cell layers thick to optimize diffusion of oxygen into our blood stream. This makes the lungs very efficient in extracting oxygen but also very vulnerable to attack. In fact, this is the Achilles' heel of the lungs. In the case of SARS, once the immune system starts attacking the infected lung cells, swelling of the lung tissues ensues. This causes fluid to build up in the lung cells called alveoli, which then prevents the oxygen from efficiently getting into the blood stream (Figure 3). In essence, the immune system, in its efforts to wipe out the virus and kill the infected cells, results in drowning on the body's own fluids. This inflammation leads to SARS (severe acute respiratory syndrome), which often results in respiratory failure and death unless supported with mechanical devices to increase delivery of oxygen, i.e., mechanical ventilators.

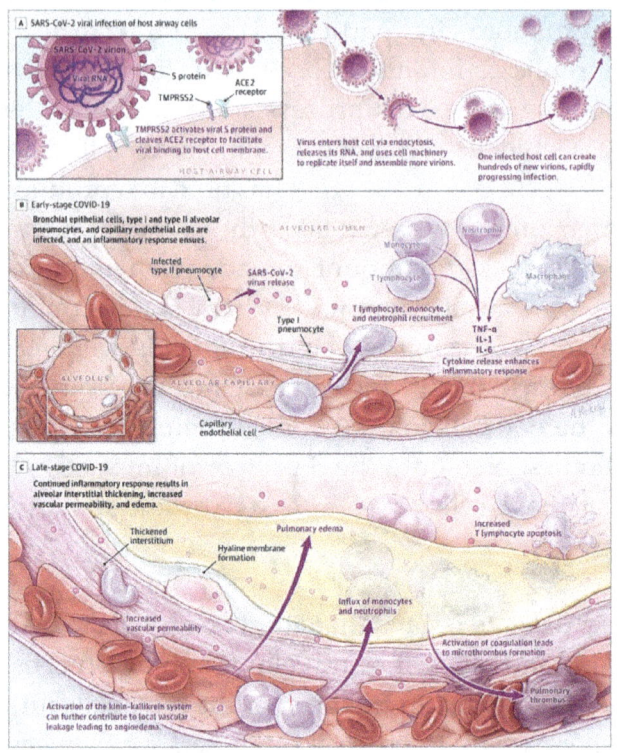

Figure. 3. (A) Covid-19 attachment, (B) Entry into Alveoli, (C) Inflammation & edema of lung tissue. JAMA 8/25/20 Vol 324, #8

Acceleration of mRNA Vaccine Development

During the outbreaks of SARS in China and Hong Kong in 2002-2003 and MERS in 2012, vaccines for these diseases were developed but never licensed. For SARS, an inactivated form of the virus was developed by Sinovac Biotech and a DNA based vaccine was developed by National Institute of Allergy and Infectious Diseases (Phase 1 trials). Most of vaccine development was halted since the virus did not re-emerge. For MERS, nine vaccines were developed, and some went into phase 1 and 2 trials. They used a live attenuated adenovirus as a vector expressing the full length of the Spike protein. These trials were also halted due to the virus not re-emerging.

Prior to COVID-19, mRNA vaccines represented 11% of all vaccines developed. The Moderna and Pfizer vaccines are the first mRNA vaccines approved for emergency use to protect against COVID-19 infections in many countries. Messenger ribonucleic acid (mRNA) vaccines are a novel technology that stimulates the body's own immune response.

In the initial phases of mRNA vaccine development, the immune system reacted to the mRNA injected into mice immunologically and neutralized the mRNA vaccine. This immune response was averted with the development of modified RNA nucleosides (building blocks of mRNA) that did not stimulate immune rejection. This scientific development was engineered and perfected by Dr. Katalin Karikó and her team.

These vaccines contain information from messenger RNA that carry the genetic code to a specific virus protein (called antigen) that allow the immune system to recognize the virus as a foreign enemy. Once the mRNA vaccine is injected into the host, mRNA transfers the information for the production of the antigen to our cell machinery that makes proteins. The information enables the body to produce a specific protein made by the virus that triggers an immune response to make neutralizing antibodies to the protein that is specific to the virus.

In the case of COVID-19, the mRNA vaccines encode for the spike (S) protein. Neutralizing antibodies targeted at the spike protein circulating in the body will immediately bind to the virus, "neutralize" it and prevent it from entering the cell and attack previously infected cells, thus protecting against infection. T cells help the immune system to fight intracellular infections and can also kill the infected cells directly (Figures 4 & 5). Thus, when a vaccinated individual comes into contact with COVID-19, the immune system will recognize the surface structure and will be able to combat the virus and eliminate it.

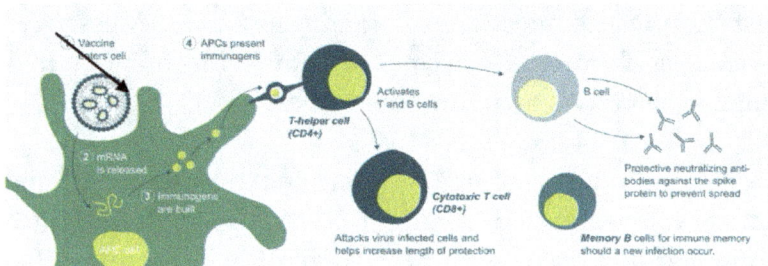

Figure 4. Credit to Biontech

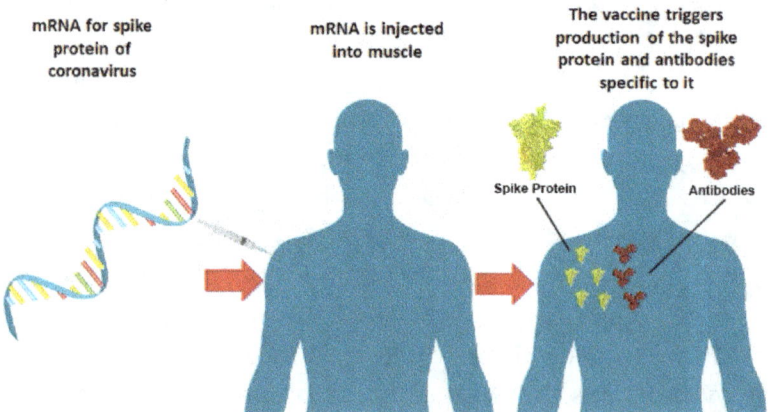

Figure 5. Credit to NIH

In contrast to conventional vaccines, a mRNA vaccine does not contain any viral proteins itself, but only the information that our own cells need to produce a virus trait that triggers the desired immune response. mRNA technology has enabled the design of multiple vaccine candidates against COVID-19.

The mRNA technology platform for developing vaccines has greatly accelerated the development of vaccines against COVID-19. Unlike traditional vaccines that require cell cultures to develop vaccines that take years to produce and test, mRNA vaccines can be immediately developed once the genetic sequence is determined and the appropriate antigen is discovered. Moderna Biotechnology developed a vaccine (mRNA-1273) in only 42 days from the date when the spike protein coding sequence was published on January 10, 2020. In total, it took one year to design, manufacture, test the safety and efficacy, and receiving approval of the vaccine.

As a result of mRNA vaccine development and due to the early manufacturing and procurement of the Pfizer and Moderna vaccines through "Operation Warp Speed", the vaccine was made available in the United States to first responders and medical professionals as of January 2021. In February 2021, vaccine was delivered to high-risk individuals and as of March 2021 it was made widely available to the public at large in the United States. As of July 2021, 65% of the United States population was vaccinated and hospitalization and deaths precipitously declined. However, in countries that lacked access to the vaccine, the case counts, and death counts continued to increase resulting in mass lock downs.

Mutations in the COVID-19 have significantly increased infection rates and death rates among younger populations. In particular, the delta variant of COVID-19 (first recognized in India in December 2020) became the dominant strain of COVID-19 globally in March 2021 and is significantly more infectious due the higher affinity of the Spike protein to "latch on" to alveoli (lung cells).

As a result, case counts began to rise globally in 2021. Fortunately, evidence shows that only a small number of patients that are vaccinated required hospitalization in the

United States and more than 90% of cases of hospitalizations and deaths due to COVID-19 resurgence were in patients that volitionally declined to get vaccinated. A study by Bernal et al. showed the current vaccines are highly effective in protecting against the delta variant (Effectiveness of Covid-19 Vaccines against the B.1.617.2 (Delta) Variant, NEJM July 2021).

With the advent of mRNA technology, vaccine development against COVID-19 has occurred at a pace previously unconceivable. This revolutionary scientific breakthrough has sharply reduced death rates globally. With increasing global mass vaccination, the spread of COVID-19 has declined, and eradication of COVID-19 is achievable and probable. This vaccine technology holds great promise in limiting and possibly averting future pandemics that most assuredly will occur.

Global Economic Implications: Increasing Wealth Gap

Historically, pandemics have always negatively affected the socio-economic activity of individuals and nation states. The societal and economic impacts are significantly more acute in the short term (years) but usually have repercussions that are seen for decades after the onset of pandemics. Currently, the societal and economic implications of COVID-19 pandemic are still evolving and the full impact at the global scale will be realized in the many years to come.

Labeled as a black swan event, the COVID-19 global recession is the deepest since WW II. The global economy contracted by 3.5% in 2020 according to the World Economic Outlook Report by the International Monetary Fund (IMF) (Figure 6). Virtually every country experienced negative growth. However, the economic impact was most pronounced in the poorest countries of the world (Noy et al. 2020) (Figure 7).

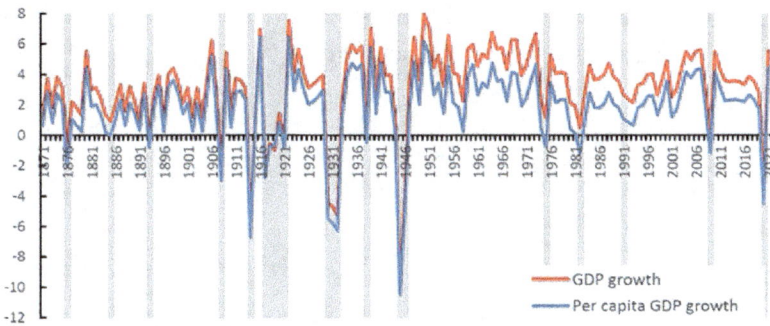

Sources: Bolt et al. (2018), Kose, Sugawara, and Terrones (2019, 2020), and IMF-WEO Apr-2021. Shaded areas refer to global recessions.

Figure 6. GDP growth & contraction 1871-2021

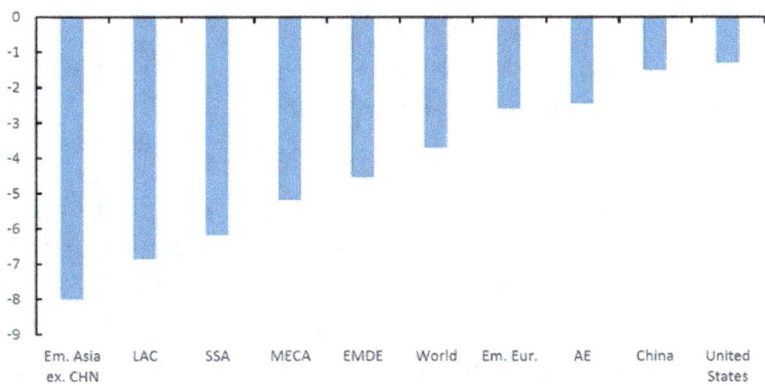

Source: IMF-WEO Apr-2021. Note: AE = Advance economies; Emerging Asia ex. CHN = emerging and developing Asia excluding China; EM. Eur = Emerging and developing Europe; LAC = Latin America and the Caribbean; MECA = Middle East and Central Asia; SSA = sub-Saharan Africa.

Figure 7. Global GDP growth 2020

While the global economy is expected to recover as society opens from lockdowns and restrictions with increased vaccination rates, the level of GDP at the end of 2021 in both advanced and developing economies is projected to remain below pre-COVID-19 levels (Figure 8). The IMF projects that in 2024 the world GDP will be 3% below the expected level had the pandemic never occurred and in low-income countries the drop would be more significant at 6%. In particular, Dijofack et al. estimate the African GDP would be permanently 1% to 4% lower had the pandemic not occurred.

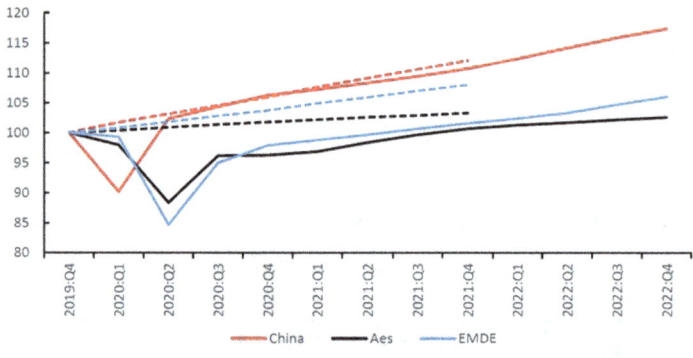

Source: IMF-WEO Jan-2021. Note: dashed lines indicated estimates from Jan-2020 World Economic Outlook Update.

Figure 8. Quarterly World GDP (GDP forecast in Jan. 2020 vs Jan. 2021, 2019)

Lower income countries experienced disproportionate economic impact from the pandemic due to poorer technological infrastructure and relative abundance of unskilled labor. Countries with higher skilled workers with access to technology allowed for workers to shift to remote work. As a result, unemployment rates in higher income countries were significantly less compared to less developed countries (Figure 9). This delay in the introduction of new technologies in developing countries has further accelerated the wealth gap as advanced economies move away from labor-intensive work force. The negative impact on labor markets and economic recovery will likely be more persistent in the developing nations due the prevalence of unskilled labor.

Figure 9. Unemployment rate globally (April 2021)

Fiscal support is another driving force of change in the global economy. Higher GDP countries provided financial relief to business and workers during the lockdowns to prevent financial collapse of their economies. This has had mixed results, particularly in the United States. Just as in all other nations with lockdowns to mitigate the spread of the virus, unemployment rose sharply in the U.S. The U. S. government pumped in significant monetary support through unemployment benefits. Paradoxically, this drove unemployment to higher levels as workers were being disincentivized to work as they received more or equal income to stay unemployed rather than being employed. This has led to a labor shortage for low skilled jobs and drove up wages due to lack of supply of labor. Just as corporations with higher income earners are shifting to remote work to minimize expenses, the pandemic is also catalyzing a shift to automation to reduce cost of labor. This will disproportionally affect lower income workers and further hasten the wealth gap (Figure 10).

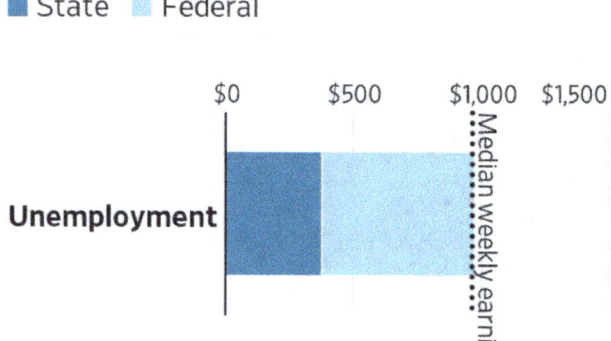

Figure 10. Federal + State unemployment benefits equaling median earnings prior to the pandemic.

Access to vaccines is another factor driving the widening wealth gap between nations. The United States, China, Russia, Canada, and European nations have been at the forefront of getting vaccines for their citizens. As a result, many of these nations have discontinued lockdowns earlier and opened their economies quicker. In contrast, developing nations and underdeveloped nations are still experiencing lockdowns and economic harm due to their inability to procure vaccines mostly due to lack of financial resources. This is most evident in India and Africa, where the failure to vaccinate its' citizens, and lack of social distancing has spiked COVID-19 infections and fatalities. This has devastated the economies in these countries and will delay economic recovery for decades.

Technological Implications: Accelerating Digital Transformation

Global lockdowns and social distancing mandates to mitigate disease spread compelled populations to shelter at home and reduce direct interactions with others. This fundamentally changed human behavior and has had many effects on the fabric of society. In particular, this accelerated the use of technology in the daily lives of citizens of all nations with a shift to remote interactions such as working, learning, shopping, and receiving medical care.

With school shutdowns, remote learning has become the primary means of education for many students. UNESCO reported on May 1, 2020, that 174 nations had school closures, affecting 1.2 billion students worldwide. UNESCO survey from 2020 reported that half of educational systems globally used some form of online learning.

At the peak of the pandemic and shutdowns in the United States, 55 million students were affected by the shutdowns. Approximately, 72% of households with students reported a shift to online education (Figure 11). Per the US Census Bureau Household Pulse Survey (Figure 12), access to online learning decreased as the household income decreased. The pandemic catalyzed an immense shift to remote learning, but also increased the educational divide between low income and middle to high-income households. This will likely manifest in lower paid jobs and less upwards mobility for students in lower income households due to lack of access to technology. On a global scale, estimated lifetime loss of labor earnings will be $10 trillion, or 12% of global GDP. This was particularly magnified in lower income families. Repercussions of these lockdowns on intellectual development will manifest years to decades after the lockdowns, exacerbating the wealth gap between high-income earners vs. lower income earners.

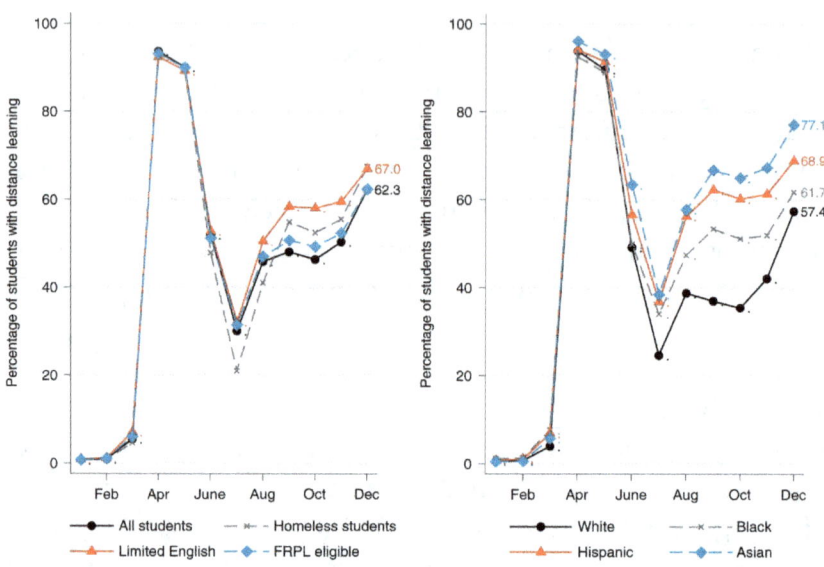

Figure 11. The sample includes 80,785 public schools with values averaged over September through December 2020.

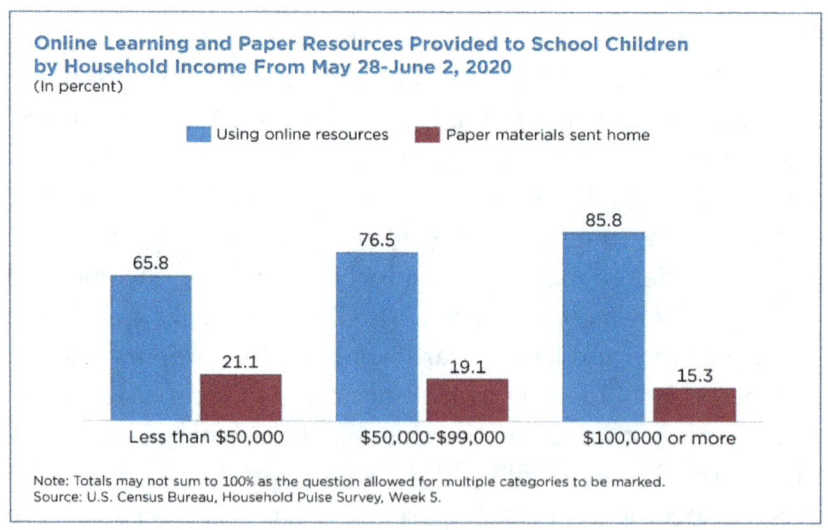

Figure 12. Online access per income of household.

In addition to the shift to remote learning, there was a parallel shift to working from home. Figure 13 shows the potential for employees to work from home in various countries based on estimates of Dingel & Neiman (2020). This figure demonstrates that the ability to work remotely increased as the GDP per capita increased for a particular nation.

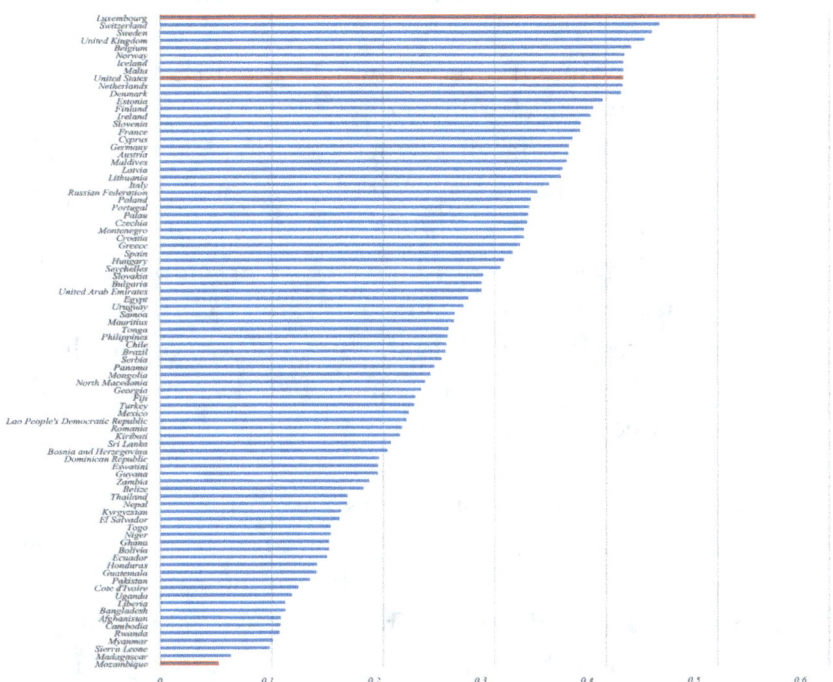

Figure 13. Share of jobs that can be done at home. *Source:* Dingel & Neiman 2020

Upwork, a firm that connects employers with freelance employees, estimates that 1 in 4 Americans (25%) of the workforce will be working remotely through 2021 and 22% (36.2 million Americans) by 2025. A survey by Owl labs found the following:

- Remote employees save an average of 40 minutes daily from commuting.

- Since 2020, people have been meeting by video calls 50% more since COVID-19.
- During COVID-19, close to 70% of full-time workers are working from home.
- After COVID-19, 92% of people surveyed expect to work from home at least 1 day per week and 80% expected to work at least 3 days from home per week.
- 23% of those surveyed would take a 10% pay cut to work from home permanently.
- People are saving on average close to $500 per month being at home during COVID-19, resulting in savings of $6000 per year.
- 81% of those surveyed believe their employer will continue to support remote work after COVID-19.
- 59% of respondents said they would be more likely to choose an employer who offered remote work compared to those who didn't.

Mercer, a human resources and workplace benefits consulting firm, surveyed 800 employers and found that 94% of the employers stated the work productivity was the same or higher since employees started working from home. Global Workplace Analytics estimates that employers can save over $11,000 dollars per year per employee. The savings are from the lower cost of office space, increased productivity, reduced absenteeism, and less turnover. They also estimate savings for employees in the $2,500 and $4,000 per year range due to lowered food costs, fuel costs, parking, and car maintenance. A survey by Slack of 9,000 workers in six countries found that 72% prefer a hybrid remote-office model with only 12% preferring to always work in an office setting (Figure 14). They also found that 13% would like to always work from home if given the choice. Another survey (Upwork) of 1,500 hiring managers found that due to COVID-19, 61.9% of the companies were planning more remote work in the post Covid-

19 years, accelerating the remote work trend that has been going on prior to COVID-19. This same report predicts 36.2 million workers or 22% of Americans will be working remotely by the year 2025. This is an 87% increase from pre-pandemic levels.

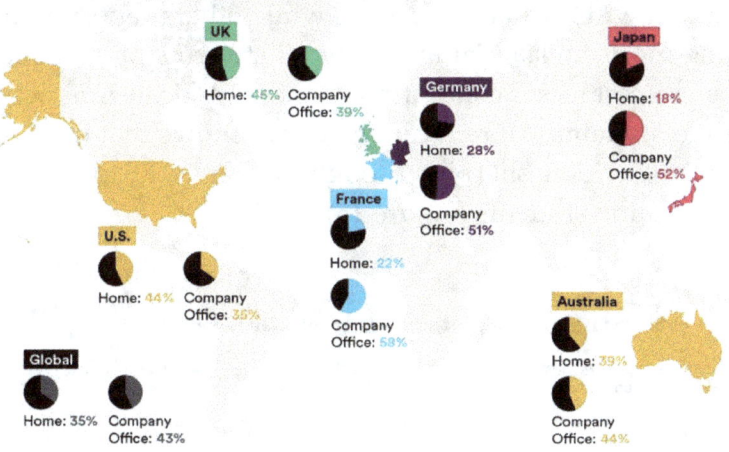

Figure 14. From Slack.com

The medical field has also experienced a significant acceleration towards digital care. In 2018, the Peterson-KFF Health System Tracker found that 2.4% of adults enrolled in large employer health plans had used telehealth services. In contrast, Medicare primary care doctor visits conducted via telehealth increased from 0.1% in February 2020 to 43.5% in April 2020. Just as the pandemic has hastened a transition to remote learning and working, it has also accelerated transformation for the medical field to increased adoption of telehealth for patient care.

Digital Divide Accelerates the Educational Divide

Emerging evidence also shows that the COVID-19 pandemic has had a strikingly negative impact on academic growth for many students of color, widening the pre-existing disparities. In fall 2020, early reports were mixed, with some finding academic growth for many students while others reporting or projecting more substantial losses for all students on average, with the greatest losses concentrated among African American and Latino students. More recent evidence shows that the gap continued to widen sharply through winter 2020 for many Black and Latino students (Figure 15).

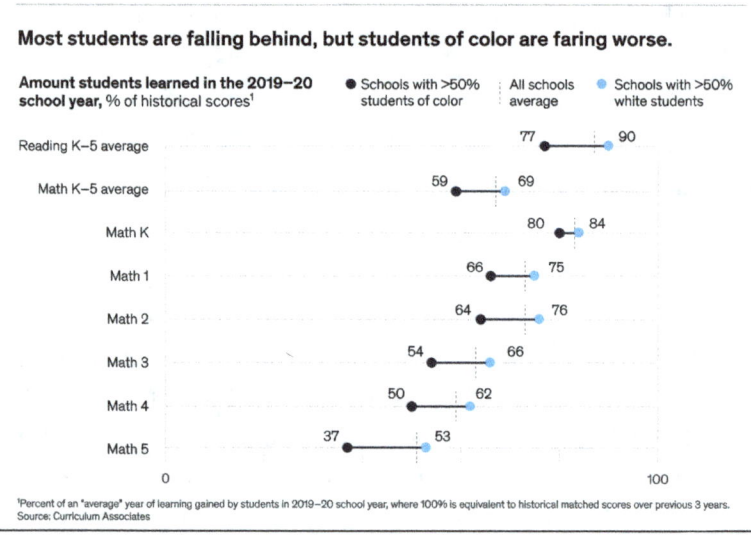

Figure 15.

A major driving force hastening the educational gap between African American and Latino students and their white and Asian counterparts is lack of access to the internet and technologies such as tablets or computers. This severely restricted live access to their classrooms and teachers (Figures 16 & 17).

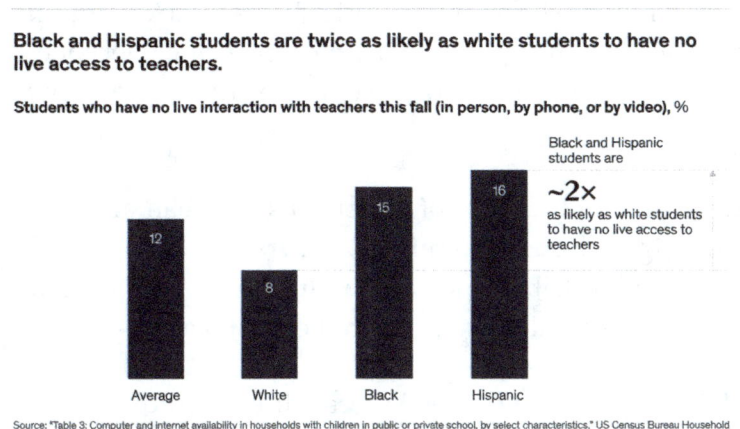

Figure 16.

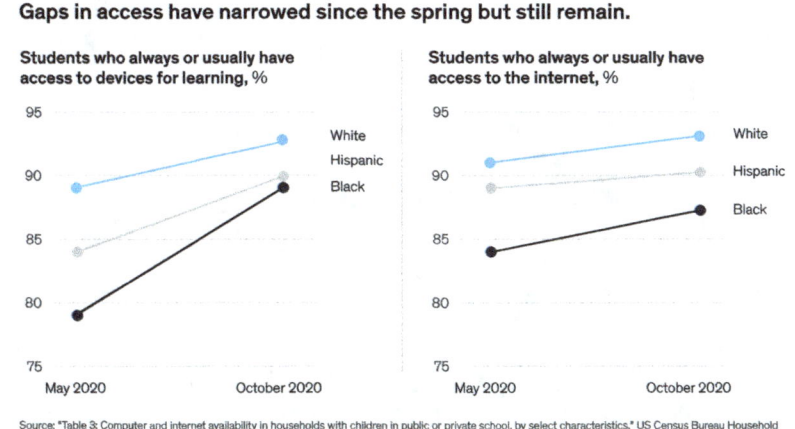

Figure 17.

Many rural and low-income communities around the world, including those in large urban areas, lack reliable, affordable internet access. More importantly, wireless technology is no longer just important for consumers and entertainment; it is rapidly becoming critical to how we connect everything in the digital world, particularly for remote learning and work. Unfortunately, the gap between the haves and have nots is widening across the US and the world due to a lack of internet. Reducing the wealth gap will depend on ensuring everyone on Earth can go wireless and access the full capability of technology.

According to the FCC, 97% of Americans in urban areas have access to a high-speed, fixed service. In rural areas, that number falls to 65% and on tribal lands to 60%. In total, nearly 30 million Americans cannot fully benefit from the digital age.

The global outlook is even more dire. According to an International Telecommunication Union report, in the developed world the internet penetration rate is 87% but just 47% in developing countries and 19% in the least developed countries (Figures 18 & 19).

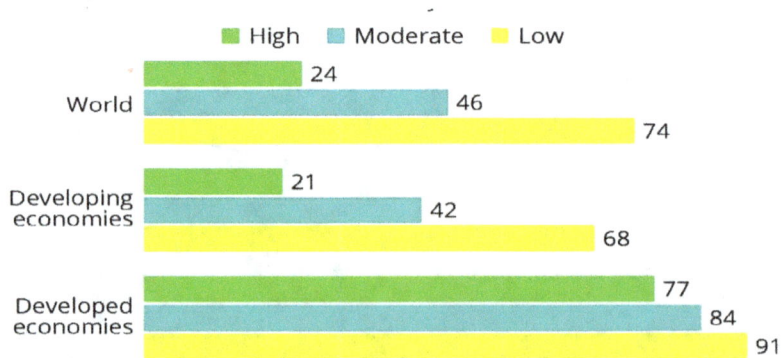

Economic vulnerability refers to times in the past year when people could/could not afford food and shelter and did/did not have friends or family to help
Source: Gallup

Figure 18. Percentage of people with internet access based on economic vulnerability (high, moderate, low vulnerability). Source Gallup.

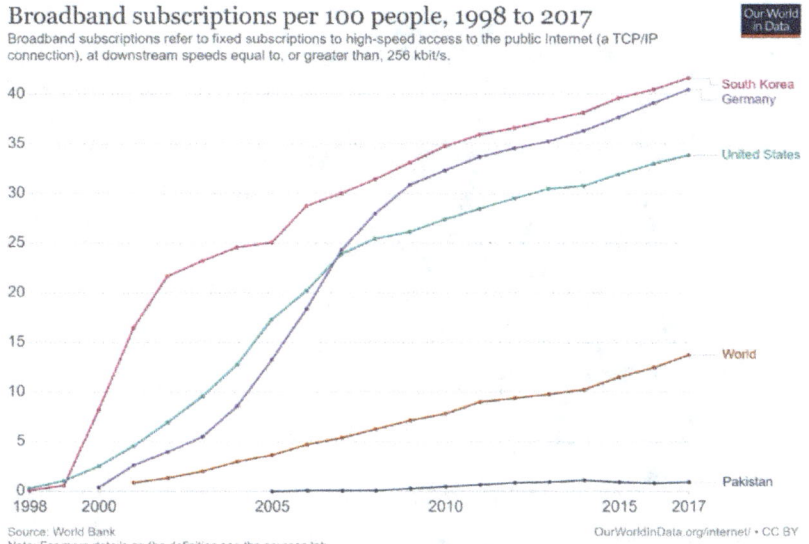

Figure 19. Source: World Bank

Remote learning during the pandemic has been far from universal and primary and secondary school students (kindergarten through 12th grade) who are defined as "poor" suffer from the digital divide disproportionately. A study by the US' National Center for Education Statistics' National Assessment of Educational Progress suggests that poor students were already less likely to have the equipment needed to attend online school prior to the pandemic; 7% of eighth graders who are poor don't have internet access compared to only 1.6% of non-poor students (Figure 20). Without reliable internet, these economic inequalities will continue to grow.

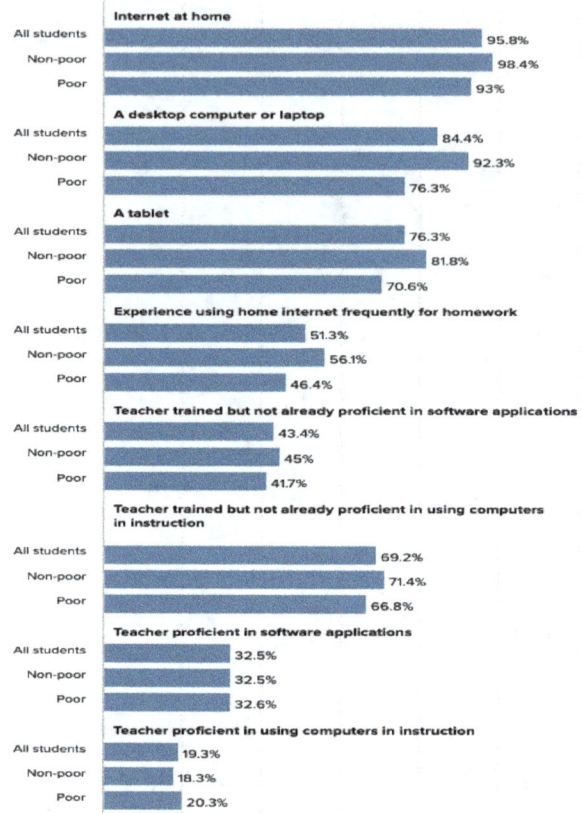

Figure 20. Share of 8th grade students with access to online learning tools by income level, 2017.

People without robust internet access are being left behind academically and economically. To close the digital divide, it is imperative to ensure everyone can be part of the "Fourth Industrial Revolution" with access to internet. The effort to bring internet to rural and poor countries is essential to bridge this gap. Most countries either lack the financial or technological means to provide internet to many of their citizens.

The current paradigm that requires optical cables and cellular towers to bring internet is beyond the capabilities of most developing nations. Rather, a fresh approach that obviates the need for this complexity and cost of infrastructure is needed. This will require low orbit satellites that can provide fast and reliable internet that does not require land-based infrastructure, but rather small satellite dishes placed in the villages or homes of rural communities.

This mission is currently underway by SpaceX that has already deployed thousands of low orbit satellites that can provide internet speeds exceeding 100mgbs (Figure 21). Starlink (subsidiary of SpaceX) has already provided thousands of individuals globally in rural areas reliable and fast internet access. Starlink will provide the ability of many students in remote areas to have access to learning tools that will help to reduce the digital divide accelerated by the pandemic.

Figure 21. Starlink low orbit satellite network

Amplification of Social Discontent and Upheaval

In an effort to mitigate COVID-19 infections, many countries introduced lockdowns to restrict the movement of people and social gatherings. The lockdowns dramatically changed the daily lives of billions of citizens. In over 180 countries, governments have implemented nationwide school closures. To measure the stringency of lockdowns, Blavatnik School of Government at the University of Oxford developed a COVID-

19 Response Tracker (OXCGRT) system with a Stringency Index ranging from 1 to 100, with 100 representing complete lockdowns. Data shows that lockdowns in the United States peaked in April 2020 with and an average index of 60 (Figure 22). Lockdowns ebbed and flowed based on surges of cases of COVID-19 after its peak in April 2020.

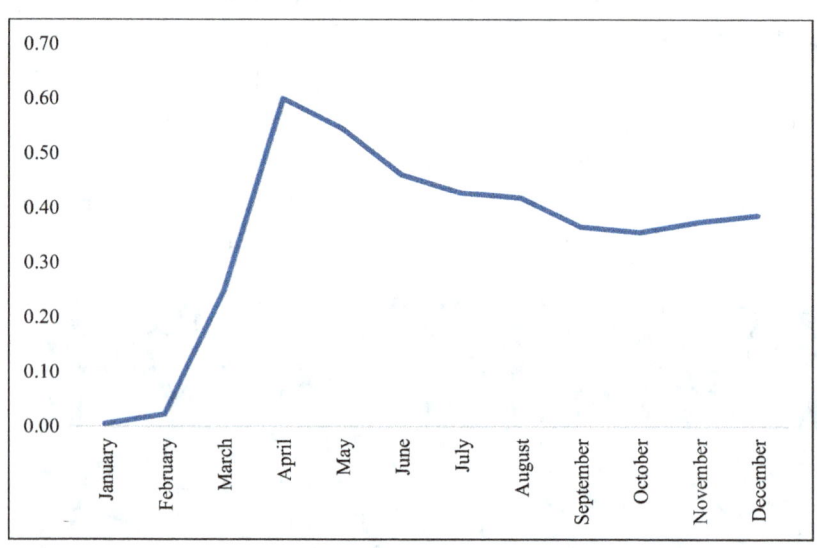

Figure 22. Average stringency index in the USA in 2020

The hallmark of the pandemic in the United States were the mass protests seen throughout the country. There is a strong correlation between lockdowns and intensity of protests in the United States. To measure this correlation, the Gini Index was developed to assess the feeling of inequality within communities through a 2019 American Community Survey using household income data. The Gini Index ranges from 0 (perfect equality) to 100 (perfect inequality). In the US, the observed index ranged from a minimum of 30 to a maximum of 71. Looking at figures 23 and 24, the greatest number of protests occurred in counties where the Gini Index was higher *pre-pandemic*.

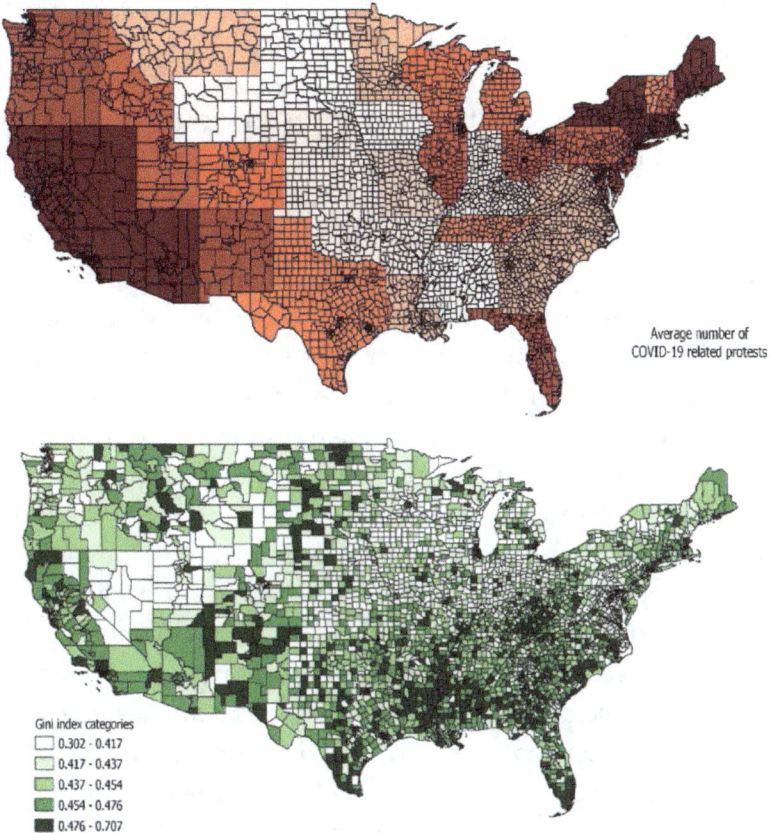

Figure 23 & 24. Note: in both maps, darker color shades represent higher intensity of the event (i.e., darker blue corresponds to stricter measures, and darker red corresponds to more protests).

These findings provide strong evidence that the implementation of lockdowns lead to the increased incidence of protests in the US counties with lower socio-economic status. Stay-at-home orders, restrictions on domestic travel, and public restrictions had the largest impact on protests. These results validate the long-standing theory that civil unrest is highest in lower income communities feeling the greatest sense of disenfranchisement. These communities already felt economically and socially oppressed before the pandemic. The imposition of restrictions further suppressed the sense of freedom within these communities and catalyzed mass protests arising from pre-existing feelings of inequality.

COVID-19 has accelerated social discontent within many countries, most profoundly in the United States with the rise in civil rights protests like Black Lives Matter, defunding the police, insurrection at Capitol Hill, Q'Anon, Proud Boys and many more. As a result, there have been attempts to make policy changes to address issues associated with income inequality. However, the pandemic has served to only highlight the inequities within the different communities in the United States without real fundamental changes on how to transform the ability of lower income communities to improve their quality of life. Unfortunately, the issue has become a political football. Hopefully, the pandemic will result in developing substantive policies or procedures to help these communities in improving educational standards and curbing violence, both of which are essential to give these communities a chance to flourish. As discussed previously, the pandemic has increased inequities between nations and within nations. If policies to address these social inequities fail to improve the socio-economic status of these communities, more protests and civil unrest will only continue to propagate and threaten the economic stability of nations.

Transformation of Societies via Pandemics

COVID-19 differs from the Black Death, the Spanish flu and smallpox in terms of disease transmission and pathophysiology. It is the least deadly and poses the lowest risk to otherwise healthy people but is the most contagious and has the highest infection rates and deaths compared to previous pandemics. The Black Death had exceedingly high case fatality rates approaching 100% for those who developed septicemia and pneumonia and 50-60% for the bubonic form of the disease. Over 33% of the European population died during the 1347 outbreak. The death rate of smallpox in Native Americans approached 50%. Spanish flu had a death rate for 2-3%. Current death rate COVID-19 measures 2.2%.

The dynamics of disease transmission have remained the same during the history of humankind. Historically, humans and domesticated animals lived in close quarters, increasing the risk for disease transmission. Communicable diseases spread more easily and rapidly where there is poverty and high population density. In India, the Spanish flu mortality rates among members of the lowest social class were three times higher than all other demographic groups. This holds true today with 73% of emerging infectious diseases in humans originating in animals, including coronaviruses.

Historically, the severity of disease determines the severity of the economic impact globally. The Black Death caused shortage of goods and labor, providing market power to the common people and sparked the European peasant revolt that resulted in permanent wage increases. Smallpox had less of a dramatic effect on the economic change, but it's devastating impact on Native American groups paved the way for European conquest of the New World, and to transfer of wealth to Europe through the mining of natural resources. This enabled the European nations to fund their empire building and colonization. Spanish flu shortly followed WWI and predominantly impacted the young to middle-aged men, which led to some labor shortage and slowed industrial growth. However, it did not cause a major decline in GDP, likely related to lower death rate. Therefore, it had less of a transformative socio-economic impact.

In contrast to these pandemics, COVID-19 poses less of a physical risk to the labor force compared to other pandemics based on its death rate. However, it has had a major economic impact from the unprecedented lockdowns. Early labor shortages were driven by shifts in demand and restrictions on movement. Mitigation efforts drove record unemployment, stock market volatility, and national GDPs to plummet. It is unclear what detrimental effects will persist in the global economy, although economists predict wage contraction and increased widespread poverty will have profound effects on emerging markets and developing economies predominantly.

Vaccine development has accelerated economic recovery of many countries with the lifting of lockdowns. However, the developing nations have had limited success in acquiring vaccines compared to developed nations. Without the vaccine, these nation states will be in a continual cycle of reopening and closing of their economies as cases rise and fall. Without mass vaccinations, underdeveloped and developing nations will fall further behind economically and intellectually, which will only exacerbate the division between poor and wealthy nations and between lower income and upper income communities.

As a result of previous pandemics, the strategic approach to minimizing deaths and socio-economic losses has shaped the global approach to COVID-19 pandemic. However, there are many lessons to be learned, in particular the need for a global approach to allow for equal and fair distribution of vaccines to reduce the tragic loss of lives due to COVID-19 and to minimize the collateral damage from lockdowns such as increased poverty, mental health related issues and the widening wealth gap. If anything is to be learned, future epidemics require international cooperation through the development of a body similar to United Nations and International Monetary Fund to provide the infrastructure and funding to build an economic reserve to help developing countries acquire and manufacture vaccines more rapidly, thereby reducing the human toll and economic impact globally. This is not only in the interest of the poorer nations but is in the best interest for all nations to promote global stability since all economies are now interdependent, as strongly evident by the collapse of global supply chains in the post COVID-19 era.

References

Glatter, MD and Paul Finkleman, History of the Plague: An Ancient Pandemic for the Age of COVID-19. American *Journal of Medicine*. 2021 Feb; 134(2): 176–181.

Bramanti B, Stenseth NC, Walløe L, Plague: a disease which changed the path of human civilization. *Adv Exp Med Biol*. 2016; 918:1–26.

Prentice MB, Rahalison L, Plague. *Lancet*. 2007; 369:1196–1207

Yang R, Plague: recognition, treatment, and prevention. *J Clin Microbiol*.

Riedel S, Plague: from natural disease to bioterrorism. *BUMC Proc*. 2005;18(2):116–124.

Demeure CE, Dussurget O, Fiol GM, *Yersinia pestis* and plague: an updated view on evolution, virulence determinants, immune subversion, vaccination, and diagnostics. *Genes Immun*. 2019;20(5):357–370.

Piret and Guy Boivin, Pandemics Throughout History Jocelyne *Front Microbiol*. 2020; 11: 631-736.

Anna Louise DesOrmeaux, "The Black Death and its Effect on 14th and 15th Century Art." *Louisiana State University*, N/A, pp. 1-22.

Cantor, N. F. *The Civilization of the Middle Ages.* Harper Perennial, 1994.

Spyrou. et al., Analysis of 3800-year-old Yersinia pestis genomes suggests Bronze Age origin for bubonic plague. *Nature Communications* 9, 2234 (2018)

Lizzie Wade, From Black Death to fatal flu, past pandemics show why people on the margins suffer most. *Science*, May 2020

Benedictow,OleJørge, *1346–1353:The Complete History of Black Death*. Cambridge, Cambridge University Press,2004

Eugenia Tognotti, Lessons from the History of Quarantine, from Plague to Influenza A. *Emerging Infectious Disease*. 2013 Feb; 19(2): 254–259.

Ainsley Hawthorn, Plague doctors and PPE: Today's gear has nothing on the medical garb of the Renaissance. Canadian Broadcasting News, August 2020

Guern & Carniel, The resurgence of the plague. *Rev Prat*. 2016 Apr;66(4):413-418.

Stefan Riedel. Edward Jenner and the history of smallpox and vaccination. *Proc (Baylor Univ Med Center)*. 2005 Jan; 18(1): 21–25.

Lakhani S. Early clinical pathologists: Edward Jenner (1749–1823). *J Clin Pathol*. 1992; 45:756–758.

Hopkins DR. Chicago: University of Chicago Press; 1983. Princes and Peasants: Smallpox in History.

Henderson et al., Working Group on Civilian Biodefense. Smallpox as a biological weapon: medical and public health management. *JAMA*. 1999; 281:2127–2137.

Littman et al., Galen and the Antonine plague. *Am J Philol.* 1973

Fenner F, Henderson DA, Arita I, et al. Smallpox and its eradication. Geneva, Switzerland: World Health Organization, 1988

Aneela Hussain, Smallpox, *Medscape,* July 2020

Joel G. Breman, D.A. Henderson, Diagnosis and Management of Smallpox. *N Engl J Med* 2002; 346:1300-1308

Nathan Nunn and Nancy Qian, The Columbian Exchange: A History of Disease, Food, and Ideas. *Journal of Economic Perspectives.* 2010; 24(2):163-188.

Beall et al. 1954, *John Hopkins University Press*

Gross et al., The Myth of the Medical Breakthrough: Smallpox, Vaccination, and Jenner Reconsidered. *J Infect Dis,* Jul-Sep 1998;3(1):54-60

N Barquet, P Domingo, Smallpox: the triumph over the most terrible of the ministers of death. *Annals of internal medicine,* 1997

N J Willis, Edward Jenner and the eradication of smallpox. *Scottish Med J,*1997 Aug;42(4):118-21

Pead et al., Benjamin Jesty: new light in the dawn of vaccination, *Lancet* 2003

The work of WHO, 1958: annual report of the Director-General to the World Health Assembly and to the United Nations

M A Strassburg, The global eradication of smallpox. *Am J Infect Control.* 1982 May;10(2):53-9.

Abbas Behbehan, The smallpox story: Life and Death of an Old Disease. *Microbiological Reviews.* Dec. 1983, p 455-509.

Barquet & Domingo, Smallpox: the triumph over the most terrible of the ministers of death. *Annals of Internal Medicine.* 1997; 127:635-642.

Farugue et al., Epidemiology, Genetics, and Ecology of Toxigenic *Vibrio cholerae.* Microbiol Mol Biol Rev. 1998 Dec; 62(4): 1301–1314.

Alwyn Collinson, How Bazalgette built London's first super-sewer. *Museum of London,* March 2019

Georges-Eugène, Baron Haussmann
French civil servant, Encyclopedia Britannica

Crystal Bennes, Eugene Belgrand & the water of Paris, *Desegno*

Lopez et al., Killed oral cholera vaccines: history, development and implementation challenges. *Ther Adv Vaccines.* 2014 Sep; 2(5): 123–136.

William B. Greenough, III, The human, societal, and scientific legacy of cholera. *J Clin Invest.* 2004 Feb 1; 113(3): 334–339.

Ali, et al., Updated global burden of cholera in endemic countries. *PLoS Negl Trop Dis.* 2015 Jun 4;9(6)

Igere and Ekundayo, Global mapping of cholera Vibrio and outbreaks in the Pre-Millennium Development Goals. *Pathog.* 2020 Dec;149

Somboonwit et al., 2017, Current views and challenges on clinical cholera. *Bioinformation*. 2017 Dec 31;13(12):405-409.

Alfani and Murphy, Plague and Lethal Epidemics in the Pre-Industrial World. *Journal of Economic History*. Mar 2017; 77(1):314-343

Huremovic et al., Brief History of Pandemics. *Psychiatry of Pandemics*, pp 7-35

Johnson and Mueller, Updating the accounts: global mortality of the 1918-1920 "Spanish" influenza pandemic. *Bull Hist Med*. Spring 2002;76(1):105-15.

Social and Economic Impacts of the 1918 Influenza Epidemic. THE DIGEST: No. 5, May 2020

Krishnamoorthy & Yoneyama, Simulating the Spread of Influenza Pandemic of 1918-1919 Considering the Effect of the First World War. May 2010

Riley et al., Nonpharmaceutical interventions implemented by US cities during the 1918-1919 influenza pandemic. *JAMA*.

Barry JM. *The great influenza: The epic story of the deadliest plague in history*. New York: Viking Press; 2004. p. 560.

Jeffery Tabuenberger, The Origin and Virulence of the 1918 "Spanish" Influenza Virus. *Proc Am Philos Soc*. 2006 Mar; 150(1): 86–112

Shope RE, Lewis PA. Swine influenza. *J Exp Med*. 1931:54.

Shope RE. The incidence of neutralizing antibodies for swine influenza virus in the sera of human beings of different ages. *J Exp Med*. 1936; 63:669–84

Jordan E. *Epidemic influenza: A survey*. Chicago: American Medical Association; 1927. p. 355.

Cox NJ, Subbarao K. Global epidemiology of influenza: past and present. *Annu Rev Med.* 2000; 51:407–21.

Simonsen L, Fukuda K, Schonberger LB, Cox NJ. The impact of influenza epidemics on hospitalizations. *J Infect Dis.* 2000; 181:831–37.

Thompson et al., Mortality associated with influenza and respiratory syncytial virus in the United States. *JAMA.* 2003; 289:179–86.

Ludwig et al., European swine virus as a possible source for the next influenza pandemic? *Virology.* 1995; 212:551–61.

Scholtissek C. Source for influenza pandemics. *Eur J Epidemiol.* 1994; 10:455–58.

Claas et al., Human influenza A H5N1 virus related to a highly pathogenic avian influenza virus. *Lancet.* 1998; 351:472–77.

Winternitz MC, Wason IM, McNamara FP. *The pathology of influenza.* New Haven: Yale University Press; 1920.

Laura Spinney, How the 1918 Flu Pandemic Revolutionized Public Health. *Smithsonian Magazine,* Sept 2017.

Taubenberger & Morens, 1918 Influenza: the Mother of all Pandemics, Emerg *Infect Dis 2006,* Jan 12(1):15-22

Aassve et al., Epidemics and trust: The case of the Spanish Flu. *Health Economics* Feb 2021, Vol 30(4):840-857

Basco et al., The redistributive effects of pandemics: evidence of the Spanish Flu. *The London School of Economics and Political Schience.* May 2020, No:308.

Becerra et al., Recent Insights into the HIV/AIDS Pandemic. *Microb Cell.* 2016 Sep 5; 3(9): 451–475.

Swanstrom R, Coffin J. HIV-1 pathogenesis: the virus. *Cold Spring Harb Perspect Med.* 2012

Campo et al., A. Oral transmission of HIV, reality or fiction? An update. *Oral Dis.* 2006;12(3):219–228

Broder S. The development of antiretroviral therapy and its impact on the HIV-1/AIDS pandemic. *Antiviral Res.* 2010;85(1):1–18.

Ford N, Calmy A, Hurst S. When to start antiretroviral therapy in resource-limited settings: a human rights analysis. *Bmc International Health and Human Rights.* 2010; 10:6.

Patteson et al., Awarness and understanding of HIV non-disclosure case law and the role of healthcare providers in discussion about the criminalization of HIV non-disclosures among women living in HIV in Canada. *Aids and Behavior.* 2020; 24:95-113.

Monshipouri, HIV/AIDS, religion, and human rights: A comparative analysis of Bangladesh, Indonesia, and Iran. *Human Rights Review.* 2012; 12:187-204.

UNAIDS. *Women and HIV – A spotlight on adolescent girls and young women*, March 2019.

Warner Greene, A history of AIDS: Looking back to see ahead. *European Journal of Immunology.* 2007;37: S94-102.

Park et al., mRNA vaccines for COVID-19: what, why and how. *International Journal of Biological Sciences.* 2021; 17(6):1446-1460.

Bashir et al., A brief review of socio-economic and environmental impact of Covid-19. *Air Quality, Atmosphere & Health.* 2020; 13:1403-1409.

Iacoella et al., Do pandemics lead to rebellion? Policy responses to COVID-19, inequality, and protests in the USA. *United Nations University* UNU-WIDER. March 2021.

Nicola et al., The socio-economic implications of the coronavirus pandemic (COVID-19): A review. *International Journal of Surgery.* 2020; 78:185-193.

Buheji et al., The extent of COVID-19 pandemic socio-economic impact on global poverty. A global integrative multidisciplinary review. *American Journal of Economics.* 2020; 10(4):213-224.

Bambra et al., The COVID-19 pandemic and health inequalities. *J Epidemiol Community Health.* 2020; 74:964-968.

Azevedo et al., Simulating the potential impacts of COVID-19 school closures on schooling and learning outcomes. *World Bank Group. Policy Research Working Paper.* June 2020: 9284

Anand & Stahel, The safety of Covid-19 mRNA vaccines: a review. *Patient Safety in Surgery.* 2021; 15:20

Wiersinga et al., Pathophysiology, transmission, diagnosis, and treatment of coronavirus disease 2019 (COVID-19) A review. *JAMA.* 2020; 324:782-793.

Fang et al., Transmission dynamics of the COVID-19 outbreak and effectiveness of government interventions: A data driven analysis. *Journal of Medical Virology*. 2020: 92(6):645-659

Yehati & Filippini. Social and economic impact of COVID-19. *Brookings Global Working Paper*. June 2021 #158.

European Centre for Disease Prevention and Control. Threat assessment brief: emergence of SARS-CoV-2 B.1.617 variants in India and situation in the EU/EEA. May 11, 2021.

Bosworth et al., Effective health communication-key factor in fighting COVID-19 pandemic. *Patient Educ Couns*. May 2020; 103(5):873-876

Matthew Rae, Potential costs of COVID-19 treatment for people with employer coverage, Peterson-KFF Health System Tracker. March 13, 2020.

www.ingramcontent.com/pod-product-compliance
Lightning Source LLC
Chambersburg PA
CBHW071511220526
45472CB00003B/978